PHYSICS IN PERSPECTIVE

STUDENT EDITION

The Nature of Physics and The Subfields of Physics

Physics Survey Committee

National Research Council

National Academy of Sciences   Washington, D.C.   1973

NOTICE: The study reported herein was undertaken under the aegis of the Committee on Science and Public Policy (COSPUP) of the National Academy of Sciences–National Research Council, with the express approval of the Governing Board of the National Research Council.

Responsibility for all aspects of this report rests with the Physics Survey Committee, to whom sincere appreciation is here expressed.

The report has not been submitted for approval to the Academy membership or to the Council but, in accordance with Academy procedures, has been reviewed and approved by the Committee on Science and Public Policy. It is being distributed, following this review, with the approval of the President of the Academy.

Available from

Printing and Publishing Office
National Academy of Sciences
2101 Constitution Avenue
Washington, D.C. 20418

LIBRARY OF CONGRESS CATALOGING IN PUBLICATION DATA

National Research Council. Physics Survey Committee. Physics in perspective.

"This book has been excerpted directly from . . . Physics in perspective, Volume I . . . [It] comprises Chapters 3 and 4–entitled 'The Nature of physics' and 'The subfields of physics,' respectively."
    1. Physics–Research–United States. 2. Science–United States. 3. National Research Council.
Physics Survey Committee. I. Title.
QC44.N37   530'.0973   73-4237
ISBN 0-309-02118-9

Printed in the United States of America

PREFACE

This book has been excerpted directly from a much more extensive report entitled *Physics in Perspective, Volume I,* the final report of the Physics Survey Committee. This excerpt is being made available to provide more convenient access to those sections of the complete report that may be of greatest interest to students and to all those wishing a relatively compact overview of the science, its frontiers, its opportunities, and its applications.

The Physics Survey Committee was appointed by the President of the National Academy of Sciences in mid-1969 and was charged with an examination of the status, opportunities, and problems of physics in the United States. Appendix A lists the members of the Committee and their affiliations. The Committee interpreted this charge broadly and attempted to place physics in perspective in U.S. society. It evolved an approach to the establishment of priorities and program emphases that may have wider potential application; it conducted detailed studies on education in physics and physics in education, on the production and utilization of physics manpower, and on the dissemination and consolidation of physics information.

The results of these studies are all included in Volume I of *Physics in Perspective*; Appendix B reproduces the table of contents of Volume I and provides a measure of its scope.

The Committee early concluded that it was essential to obtain detailed information from experts in each of a number of physics subfields and interface areas. For each of these subfields and interface areas a panel of recognized experts was brought together to survey and report on their respective subject areas. Appendix C lists these panels together with their chairmen.

Several of the subfields have relatively well-defined and traditional boundaries in physics. Included are the core subfields of acoustics, optics, condensed matter, plasma and fluids, atomic, molecular, and electron physics, nuclear physics, and elementary-particle physics. The reports of these panels constitute *Physics in Perspective, Volume II, Part A*. In addition, there are several important interface areas between physics and other sciences. In the case of astronomy, where activity is particularly vigorous at the interface and is overlapping, the Physics and Astronomy Survey committees formed a joint panel that reported on astrophysics and relativity, an area of special interest to both. The broad area in which physics overlaps geology, oceanography, terrestrial and planetary atmospheric studies, and other environmental sciences was defined as earth and planetary physics, and a panel was established to survey it. In covering the physics–chemistry and physics–biology interfaces, the broader designations "physics in chemistry" and "physics in biology" were chosen to avoid restricting the work of the panels to the already traditional boundaries of these interdisciplinary areas. The reports of all these panels, together with those considering education, instrumentation, and information dissemination and consolidation constitute Volume II, Part B.

A number of subjects in classical physics—such as mechanics, heat, thermodynamics, and some elements of statistical physics—were not considered explicitly in the survey. This omission is in no sense intended to imply any lack of their importance but merely indicates that they are mature subfields in which relatively little research per se is currently being conducted.

The panel reports have been published in Volume II in the form submitted to the Committee, not only to provide the detailed technical background and documentation for many of the Committee's findings but also because they provide to a unique degree a measure of the vitality and strength of the different subfields of physics. Repeatedly in its activity the Committee has been reminded of the unity of physics and, indeed, of all science. This intellectual thread is interwoven through all the panel reports.

This book comprises Chapters 3 and 4—entitled "The Nature of Physics" and "The Subfields of Physics," respectively—from *Physics in Perspective, Volume I*. Chapter 3 presents a discussion in some depth of the nature of physics as a science and as a part of Western culture. It addresses such basic questions as: What is physics? Why are physicists interested in it? Why should anyone else be interested in it? The different sections of Chapter 4 are in many ways summaries of the corresponding panel reports. Each summary describes the present status, recent developments and achievements, and outstanding opportunities now identified, together with illustrations selected to demonstrate the vital unity and coherence of physics. This unity is the subject of the concluding section of Chapter 4, in which is illustrated first the remarkable similarity of concepts and theoretical techniques that are employed in condensed matter, atomic and molecular, nuclear, and elementary-particle physics and then the even more remarkable extent to which almost every branch of classical and modern physics contributes to the understanding of the recent beautiful measurements on pulsars. In considering achievements and opportunities, attention was directed to those relating to the solution of major national problems and to other fields of science and technology as well as those internal to physics.

To facilitate a more detailed examination of each subfield, we have divided each into what we have defined as program elements. These are scientific subgroupings having reasonably identifiable and unambiguous boundaries with which it is possible, with reasonable accuracy, to associate certain fractions of the total manpower and federal funding in each subfield. Discussion of these program elements appears in Appendix 4.A; Appendix 4.B discusses recent data on migration of physicists between different areas of specialization, while Appendix 4.C presents a brief listing of suggested supplementary references for further reading in each of the subfields.

To make this material available to the widest possible audience, and most particularly to students, we have minimized production costs through direct reprinting of the Volume I chapters, without change in pagination and with black and white reproductions of all color plates in the original.

The Committee has acted under the aegis of the National Academy of Sciences Committee on Science and Public Policy. Its work has been supported equally by the Atomic Energy Commission, the Department of Defense, the National Aeronautics and Space Administration, and the National Science Foundation; it also has received important support from the American Physical Society and from the American Institute of Physics.

We cannot hope to acknowledge, in detail, all the assistance that we have received from many persons and organizations throughout the country during the course of this survey. We are particularly grateful to George W. Wood, Charles K. Reed, and Bertita Compton, without whose continuing assistance the survey would never have been completed. And we would be remiss, indeed, if we did not especially recognize the work, far beyond any call of duty, of Beatrice Bretzfield, who acted as secretary to the Physics Survey at the National Academy of Sciences since 1970, and of Mary Anne Thomson, my administrative assistant at Yale.

D. ALLAN BROMLEY, *Chairman*
Physics Survey Committee

New Haven, Connecticut
January 26, 1973

CONTENTS

* Page numbers referred to here are new for this edition and appear at the bottom
of the page in square brackets. Page numbers appearing at the top of the page are
from *Physics in Perspective, Volume I,* and have been left intact from the original
printing.

3
THE
NATURE
OF
PHYSICS

scire—to know
scientia—knowledge

Nam et ipsa scientia potestas est

FRANCIS BACON (1561–1626)
Of Heresies

INTRODUCTION

Science is knowing. What man knows about inanimate nature is physics, or, rather, the most lasting and universal things that he knows make up physics. Some aspects of nature are neither universal nor permanent—the shape of Cape Cod or even a spiral arm of a galaxy. But the forces that created both Cape Cod and the spiral arm of stars and dust obey universal laws. Discovering that has enabled man to understand more of what goes on in his universe. As he gains more knowledge, what would have appeared complicated or capricious can be seen as essentially simple and in a deep sense orderly. The explorations of physical science have brought this insight and are extending it—not only insight but power. For, to understand how things work is to see how, within environmental constraints and the limitations of wisdom, better to accommodate nature to man and man to nature.

These are familiar and obvious generalities, but we have to begin there

(*Above*) Hurricane Gladys was stalled west of Naples, Florida, when photographed from Apollo 7 on October 17, 1968. Its spiraling cumuliform-cloud bands sprawled over hundreds of square miles. A vigorous updraft hid the eye of the storm by flattening the cloudtops against the cold, stable air of the tropopause (then at 54,000 feet) and forming a pancake of cirrostratus 10 to 12 miles wide. Maximum winds near the center were then 65 knots. [From National Aeronautics and Space Administration, *This Island Earth*, O. W. Nicks, ed., NASA SP-250 (U.S. Government Printing Office, Washington, D.C., 1970).]

(*Below*) NGC 3031, spiral nebula in Ursa Major photographed with the 200-inch reflector of the Palomar Observatory. [Courtesy Hale Observatories]

if we want to discuss the value of physics in today's and tomorrow's world. Going beyond generalities evokes sharp questions from several sides. Will the knowledge physicists are now striving to acquire have intrinsic value to man, whether it has practical application or not? Is it possible to promise that material benefits will eventually accrue, at least indirectly, from most of the discoveries in physics? How does technology depend on further advances in physics (and vice versa)? How does physics influence other sciences? Is vigorous pursuit of new knowledge in physics still beneficial, as it demonstrably was in the past, to chemistry, astronomy, and the other sciences for which physics provided the base? Is physics perhaps approaching the end of its mission, without very much more to discover? Has the physicist himself an intrinsic value to human society? Must he justify his work by relating it to pressing social problems? Only one person in several thousand is a physicist; will it matter to the others what he does, or that he is there at all? Or is that the right test to apply, no matter how it comes out?

We speak briefly to such questions in this and the following chapter. The entire Report, including the reports of the various subfield panels, provides in copious detail answers to some of these questions or facts from which a reader can form his own judgment, for these are not questions that even all physicists would answer in just the same way.

FUNDAMENTAL KNOWLEDGE IN PHYSICS

Mathematics deals with questions that can be answered by thought and only by thought. A mathematical discovery has a permanent and universal validity; the worst fate that can overtake it is to be rendered uninteresting or trivial by enclosure within a more comprehensive structure. Mathematicians make up, or one could say discover, their own questions in the timeless universe of logical connection. In a science such as geology, on the other hand, the questions arise from local, more or less accidental features of nature. How was this mountain range formed? Where was Antarctica two billion years ago? To answer such questions one has to sift physical evidence. The answers are not universal truths. Geology, as its name attests, differs from planet to planet.

Physics, like geology, is concerned with questions that cannot be decided by thought alone. Answers have to be sought and ideas tested by experiment. In fact, the questions are often generated by experimental discovery. But there is every reason to believe that the answers, once found, have a permanent and universal validity. All the evidence indicates that physics is essentially the same everywhere in the visible universe. A physicist who

[3]

asks, "Does the neutron have an electric dipole moment?" and turns to experiment to find out, could as well perform the experiment on any planet in any galaxy—it is just more convenient here at home. The question itself concerns a fact as general as (and perhaps even more basic than) the size of the universe. Physics is the only science that puts such fundamental questions to nature.

Take the question: Do all electromagnetic waves, including radio waves and light waves, travel through empty space at the same speed? Present theories assume so, but the contrary is at least conceivable. Perhaps there is a difference so slight that it has not been noticed. To decide, the physicist must turn to experiment and observation. In fact, this particular question has recently received renewed attention. As not infrequently happens, the most sensitive test was applied by asking what sounds like a different question but can be shown to be logically equivalent. Indeed, the experimental evidence shows that the speed of long and short electromagnetic waves is the same to extraordinarily high precision. The result implies that the light quantum, the photon, cannot have an intrinsic mass as great as 10^{-20} * of the mass of an electron. No one was astonished by the result. Most physicists have always assumed the photon rest mass to be exactly zero and can only be relieved that such peculiarly perfect simplicity has survived closer scrutiny. Those who examined the evidence may have been a little disappointed—but their time was not wasted.

This quite unsensational episode is characteristic of fundamental inquiries in several ways. Fundamental experiments in physics often—indeed usually—yield no surprises. However, had the result been otherwise, it would not have demolished electromagnetic theory. A generalization or enlargement of the theory would have been necessary. Finally, the test, sensitive as it was, could not settle the question once and for all, for no real experiment achieves infinite precision. So the question will doubtless be raised again, in one form or another, should a new experimental technique or a bright idea create the opportunity for a significantly more stringent test. An equally fundamental assumption, the proportionality of inertial and gravitational mass, was tested in experiments of successively higher precision by Newton (1686), Bessel (1823), Eötvös (1922), and Dicke (1964) —in the last case to an accuracy of 10^{-11}.

Thanks to such relentless probing of its foundations, even where they appear comfortably secure, physics has acquired a base far more solid than is popularly appreciated. When a physicist states that a proton carries

* Physicists commonly use powers of 10 as a convenient shorthand for expressing large numbers. Thus, for example, 1000 becomes 10^3 and 1,000,000 becomes 10^6. The numeral 1 followed by zeros equal in number to the value of the exponent is the rule of thumb. This notation is used throughout the Report.

[4]

a charge equal to that of the electron, he can point to an experiment that proved any inequality to be less than one part in 10^{20}. When he expresses confidence in the special theory of relativity, he can refer to a multitude of experiments under ultrarelativistic conditions in which even a slight failure of the theory would have been conspicuous. Electromagnetic interactions are today more completely accounted for—that is to say, better understood—than any other phenomena in physics. Quantum electrodynamics, the modern formulation of electromagnetic theory, has now been tested experimentally over a range of distance from 10^9 cm down to 10^{-14} cm, a range of 10^{23}. This theory was itself developed in response to experiments that revealed small discrepancies in the predictions of the much less complete theory that preceded it. No one will be much surprised if quantum electrodynamics in its present form fails to work for phenomena involving still smaller distances; that will not diminish its glory or its validity within the vast range over which it has been tested. Nor did the extension of electromagnetic theory into quantum electrodynamics deny the essential truth of Maxwell's equations for the electromagnetic field. Knowledge thus won is about as permanent an asset as mankind can acquire.

The most fundamental aspects of the physical universe are manifest in symmetries. In the history of modern physics, the concept of symmetry has steadily become more prominent. The beautiful geometrical symmetry of natural crystals was the first evidence of the orderliness of their internal structure. Exploration of the arrangement of atoms in crystals by x-ray diffraction, begun 60 years ago, has mapped the structure of thousands of substances and is now revealing in detail the architecture of the giant molecules involved in life. Meanwhile, physicists became concerned with more than just geometrical symmetry. Symmetry, in the broadest sense, involves perfect indifference. For example, if two particles are distinguishably different in some ways but show absolutely the same behavior with respect to some other property, a physicist speaks of symmetry. The notion of identity of particles is intimately related. All these ideas acquire their real importance in quantum physics, where an object, a molecule for instance, is completely characterized by a finite number of attributes.

In probing questions of symmetry in the domain of elementary particles, the physicist is again, like the first crystallographers, seeking a pattern of all-pervading order. A sobering lesson learned from modern particle physics, a lesson the Greek atomistic philosophers would have found unpalatable, is that man is not wise enough to deduce the underlying symmetries in nature from general principles. He has to discover them by experiment and be prepared for surprises. No one guessed before 1956 that left and right made a difference in the interaction of elementary particles. After it was found that the true and perfect indifference in the weak interactions is not

[5]

left/right but left-electron/right-positron, it was again disconcerting to find even this rule of symmetry violated in certain other interactions. But it has by no means been all surprises. Symmetry rules guessed from scattered clues often have been amply corroborated by later experiments; and in particle physics, thinking about symmetries has been enormously fruitful. A grand pattern is emerging, largely describable in terms of symmetries, that makes satisfying sense.

The primary goal of research in fundamental physics is to understand the interactions of the very simplest things in nature. That is a basis, obviously necessary, for understanding larger and more complicated organizations of matter anywhere in the universe. The shape of a particular galaxy is not, from this rather narrow point of view, a fundamental aspect of nature, but the motion of an electron in a magnetic field is fundamental and has implications for many things, including life on earth and the shapes of galaxies.

However, physics has to be concerned with more than the elementary few-body interactions of particles and fields. It can be a gigantic step from an understanding of the parts to an understanding of the whole. To appreciate the total task of physics, a broader view of what is fundamental is necessary. Consider, for example, man's practically complete ignorance of the evolution of the flat, patchily spiral distribution of gas and stars that he calls his own (Milky Way) galaxy. (Only very recently some plausible theories have been developing; it is too early to say how much they can explain.) The interaction of molecules, atoms, ions, and fields is now well enough known for this problem, and Newtonian gravitation, on this scale, is unquestionably reliable. With these simple ingredients, why doesn't the problem reduce to a mere mathematical exercise? One good reason— perhaps not the only reason—is that a complete and general theory of turbulence is lacking. It is not just a lack of efficient methods of calculation. There is a gap in man's understanding of physical processes, which remains unclosed, even after the work of many mathematical physicists of great power. This gap is blocking progress on several fronts. When a general theory of turbulence is finally completed, which probably will depend on the work of many physicists and mathematicians, a significant permanent increase in man's understanding will have been achieved. That will be fundamental physics, using both fundamental and physics in a broad sense, as, to take an example from the recent past, was the explanation of the mystery of superconductivity by Bardeen, Cooper, and Schrieffer. Although pedantically classifiable as an application of well-known laws of quantum mechanics, it was truly a step up to a new level of understanding.

These two examples, turbulence and superconductivity, stand near opposite boundaries of a wide class of physical phenomena in which the behavior of a system of many parts, although unquestionably determined by the interaction of the elementary pieces, is not readily deducible from them. Physicists know how to deal with total chaos—disorganized complexity. Statistical mechanics can predict anything one might want to know about a cubic centimeter of hydrogen gas with its 10^{19} molecular parts. As for the hydrogen molecule itself, it presents the essence of ordered simplicity. Its structure is completely understood, its properties calculable by quantum mechanics to any desired precision. What gives physicists trouble, to continue the classification suggested by Warren Weaver, is organized complexity, which is already present in a mild form in so familiar a phenomenon as the freezing of a liquid—a change from a largely disordered to a highly ordered state. Here a general feature is that what one molecule prefers to do depends on what its neighbors are already doing. How drastically that feedback changes the problem is suggested by the lack of a theory that can predict accurately the freezing point of a simple liquid. In the physics of condensed matter, many such problems involving cooperative phenomena remain to challenge future physicists. A turbulent fluid, on the other hand, confronts the physicist with a system in which order and disorder are somehow blended. Complex it certainly is, but not wholly disorderly, admitting no clean division between the random flight of a molecule of the fluid and the organized motion of a row of eddies.

The solution of these major problems of organized (or partly organized) complexity is absolutely necessary for a full understanding of physical phenomena. Extraordinary insight and originality will surely be needed, as indeed they always have been. The intellectual challenge is as formidable as that faced by Boltzmann and Gibbs in the development of statistical mechanics. The consequences for science of eventual success could be as far-reaching.

Broadly speaking then, the unfinished search for fundamental knowledge in physics concerns questions of two kinds. There are the primary relations at the bottom of the whole structure. How many remain to be discovered and how small the number to which they can ultimately be reduced are not yet known. Then there is the knowledge needed to understand all the behavior of the aggregations of particles that make up matter in bulk. Here the mysteries are perhaps not so deep, although the remaining unsolved problems are of formidable and subtle difficulty. It is easier to imagine how this part of the development of fundamental physics could be concluded, if the even more difficult problem of organized complexity in living organisms is left for the physiologist, assisted by the biochemist and biophysicist, to solve.

What has been learned in physics stays learned. People talk about scientific revolutions. The social and political connotations of revolution

[7]

evoke a picture of a body of doctrine being rejected, to be replaced by another equally vulnerable to refutation. It is not like that at all. The history of physics has seen profound changes indeed in the way that physicists have thought about fundamental questions. But each change was a widening of vision, an accession of insight and understanding. The introduction, one might say the recognition, by man (led by Einstein) of relativity in the first decade of this century and the formulation of quantum mechanics in the third decade are such landmarks. The only intellectual casualty attending the discovery of quantum mechanics was the unmourned demise of the patchwork quantum theory with which certain experimental facts had been stubbornly refusing to agree. As a scientist, or as any thinking person with curiosity about the basic workings of nature, the reaction to quantum mechanics would have to be: "Ah! So that's the way it really is!" There is no good analogy to the advent of quantum mechanics, but if a political–social analogy is to be made, it is not a revolution but the discovery of the New World.

THE QUESTION OF VALUE

Most people will concede that fundamental scientific knowledge is worth its cost if it contributes to human welfare by, even indirectly, promoting the advance of technology or medicine. It is easy to support a claim for much of physics. But now that some frontiers of fundamental research have been pushed well beyond the domain of even nuclear engineering, that justification is not always plain to see. A connection between many-body theory and the latest semiconductor device is not much more difficult to trace than the connection between thermodynamics and a jet engine. But it is not easy to foresee practical applications of the fundamental knowledge gained from very-high-energy experiments or, say, tests of general relativity.

Two responses can be made, each of which has some validity. First, inability to foresee a specific practical application does not prove that there will be none. On the contrary, that there almost certainly will be one has become a tenet of conventional wisdom, bolstered by familiar examples such as Rutherford's denial of the possibility of using the energy of the nucleus. Applying this principle to strange-particle interactions will probably raise fewer doubts among laymen than among physicists. Even here the conventional wisdom may be sound after all. High-energy physics is uncovering a whole new class of phenomena, a "fourth spectroscopy" as it is termed elsewhere in this Report. In the present state of ignorance, it would be as presumptuous to dismiss the possibility of useful application as it would be irresponsible to guarantee it.

[8]

A secondary benefit that can be expected, as a return for supporting such research, is the innovation and improvement in scientific instrumentation that such advanced experiments stimulate. (This point is discussed in a subsequent section of this chapter on the contributions of physics to technology.) Other sciences also benefit from the development of experimental techniques in physics.

But these responses do not squarely face the question: What is fundamental knowledge itself worth to society? Elementary-particle physics provides an example. A permanent addition to physicists' knowledge of nature was the recognition, several years ago, that there are two kinds of neutrino. This fact, although compatible with then existing theory, was not predictable *a priori,* nor is the reason understood. The question was put to nature in a fairly elaborate high-energy experiment, at a total monetary cost that reasonable accounting might put at $400,000 (not including beam time on the Alternating Gradient Synchrotron Accelerator). The answer was unequivocal: The electron neutrino and the muon neutrino are not identical particles.

For physics this was a discovery of profound significance. Neutrinos are the massless neutral members of the light-particle or lepton family, of which the familiar electron and its heavier relative, the muon, are the only other known members. Just how these particles are related—even why there is a muon—is one of the central puzzles of fundamental physics, a puzzle that is as yet far from solution. Obviously, it was not about to be solved while physics remained ignorant of the fact that there are two kinds of neutrino, not just one.

Still, how does this bit of knowledge benefit the general public, interesting as it may be to the tiny fraction of scientists who know what "two kinds" means in this connection? The answer must be that the discovery was a step—a necessary step—toward making nature comprehensible to man. If man is going to understand nature, he has to find out how it really is. There is only one way to find out: Experiment and observe. If man does not fully understand the leptons, he cannot claim to understand nature.

On the other hand, the neutrino is a rather esoteric creature. It would be absurd to expect wide and instant appreciation of this fundamental discovery. Even of fundamental knowledge there is too much for most people to absorb. Many a physicist who could calculate on the back of an envelope the neutrino flux from the sun remains complacently ignorant of the location and function of the pituitary gland in his body. The point is that the value of new fundamental knowledge must not be measured by the number of people prepared to comprehend it. To say that man understands this or that aspect of nature usually means that some people do, and that they understand it sufficiently well to teach it to any who care to

[9]

learn and to maintain a reliable base from which they or others can explore still further. The great thing about fundamental scientific knowledge is that it is an indestructible public resource, understandable and usable by anyone who makes the effort. When so used in its own domain, it is a thing of beauty and power.

The great American physicist Henry Rowland once replied to a student who had the temerity to ask him whether he understood the workings of the complicated electrostatic machine he had been using in a demonstration lecture: "No, but I could if I wanted to." Knowledge of fundamental scientific laws makes for economy of human thought. It is the great simplifier in a universe of otherwise bewildering complexity. It is not necessary to analyze every cogwheel in an alleged perpetual motion machine to know that it will not work or to keep tracking all the planets to be sure that they are not about to collide. The revelation that the electron and muon neutrinos are different, although it might appear to have complicated matters, was in fact a step toward ultimate simplicity, because it brought closer the essential truth about leptons.

Some of the fundamental ideas of physics have slowly become part of the mental furnishings of most educated people. The following statements probably would elicit general assent: All substances are composed of atoms and molecules; nothing travels faster than light; the universe is much larger than the solar system and much older than human life; energy cannot be obtained from nothing, but mass can be turned into energy; motions of planets and satellites obey laws of mechanics and gravity and can be predicted precisely. That is surely a rather meager assortment, but, even so, what an immense difference there is between knowing these few things and not knowing them—a difference in the relation of a person to his world. A child asks his father "What is a star?" or "How old is the world?" In this century he can be answered, thanks to hard-won fundamental knowledge. What is that worth? The answers can hardly contribute to anyone's material well-being, present or future. But they do enlarge the territory of the human mind.

Much of what modern physics has learned has not yet become common knowledge. Here is an example. Not only physicists but everyone who has studied quantum mechanics knows that all known particles, without exception, fall into one or the other of just two classes, called fermions and bosons, which differ from one another profoundly on a certain question of symmetry. The difference is as fundamental as any difference could be. Although usually expressed somewhat abstractly, the distinction is less recondite than some theological distinctions over which men have quarreled fiercely. Its concrete manifestations are vast, among them the astonishing properties of superfluid helium (a boson liquid), the electrical properties of metals, and, indeed, through the Pauli exclusion principle, the very existence of atoms and molecules, hence of life. Now it would seem that this profound, essentially simple truth about the physical universe ought to be known to most fairly well-educated persons, to as many, perhaps, as understand the difference between rational and irrational numbers. Yet, it is probably safe to say that a majority of college graduates have never heard of fermions and bosons, and that an even larger majority is not equipped to understand what the distinction means. Probably far less than 10 percent of current college graduates have had a course in physics or chemistry in which the exclusion principle was mentioned. Perhaps 10 percent will learn enough mathematics so that, if they are interested, they could be made to understand a statement such as "the wavefunction changes sign on exchange of particles."

But can anyone except physical scientists be interested in such a question? History suggests that it is possible. Long before the atomic bomb made mc^2 a catchword, the theory of relativity (both special and general) engaged the public interest more intensely than anything else in twentieth century physics. The fascination lay not only in the enigmatic figure of Einstein and the notion of a theory that, as the newspapers were fond of claiming (quite erroneously), only 12 men could understand. There was at the same time a sustained, genuine intellectual interest, at all levels of understanding commencing with zero, in the puzzling implications of new ideas about space and time. To this day, nothing beats the twin paradox for stirring up spirited argument in an elementary physics class. People who have any interest at all in ideas seem to be more interested, on the whole, in fundamental questions than in practical questions. It is usually easier to interest an intelligent layman in the uncertainty principle than in how the mass spectrograph works. Of course, that is true only if he or she can be given some idea, not wholly superficial, of the meaning of the uncertainty principle. This can be done; it has been accomplished many times, in different styles, by imaginative teachers and writers. Nor is it hard to convey to a thoughtful person the essential notion of antimatter, or the question of left- and right-handedness in nature, both ideas that intrigue many nonscientists. It may even turn out that nonphysicists of the next generation, many of whom were brought up with the new math and are on speaking terms with computers, will find the abstract rules of particle physics a more satisfying statement about nature than would an old-fashioned physicist.

Admittedly, there are difficulties in engaging the active interest of nonscientists in some of the most fundamental ideas of physics. They are illustrated in the example of the fermion–boson distinction. Unlike relativity, this subject makes no connection with familiar concepts such as time, space, and speed. No paradox or controversy stirs the imagination in first acquaintance. The intelligent layman can only listen to the explanation of fermions and bosons as if he were hearing a story about another world. There is nothing to argue about. It may serve as brief intellectual entertainment; most likely it will not impinge on or disturb the ideas he already has. He may not be eager to tell someone else about it. In that case it can hardly be claimed that the person has gained something of permanent value to him. And yet, when followed to a slightly deeper level, this idea has a direct bearing on a question that has engaged human throught for 2000 years—the ultimate nature of substance. It gives a most extraordinary answer to philosophical questions about identity of elementary particles, questions that were already implicit in the cosmology of Democritus but were never faced before quantum mechanics. Here, too, is a key to the wave–particle duality with which the quantum world confounded man's mechanistic preconceptions. The philosophical implications of the fermion–boson dichotomy are still, after 40 years, poorly understood by philosophers.

So the problem is one of teaching. Very many people who are not scientists are interested or can be interested in the basic questions that have always attracted human curiosity. The discoveries of physics, even those presently described in abstruse language, bear directly on some of these questions—so directly that when understood they can transform a person's conception of the atomic world or the cosmos. To promote that understanding is a task for the scientist as teacher, in the broadest sense of teacher. In the short run, drawing a potential audience from college graduates of the past 20 or 30 years, and perhaps the next 10, the physicist must apply his imagination and ingenuity to convey interesting and meaningful, and essentially true, accounts of some of the fundamental developments in physics. It is to be hoped that some day the educated layman he addresses will have had enough physical sciences and mathematics in his general education to turn a discussion of the symmetries of elementary particles into some sort of dialogue.

The audience need not be of a size that would impress a national advertiser but only a few million people—a few hundred, say, for every physicist. Of course, the distribution of potential interest and comprehension is a many-dimensional continuum. Everyone ought to be, and can be, given some glimpse of what fundamental physics is about. However, it is impossible to compare the value of a brief exposure of 10^8 people to news of a discovery in physics with the value of sustained and active interest on the part of 10^6 people. Both are valuable now, and both will help, in the long run, to make the fundamental knowledge that physics is securing meaningful and useful to all people.

The value of new fundamental scientific knowledge is not, after all, contingent on its appreciation by contemporary society. It really does not matter now whether Clerk Maxwell's ideas were widely appreciated in Victorian England. Their reception is interesting to the historian of science, but mainly as a reflection of the attitudes and structure of the society in which Maxwell worked. The full value of a scientific discovery is concealed in its future. But even as the future unfolds, the value that one may set on an isolated piece of fundamental knowledge often becomes uncertain because of the interconnection in the growing structure. In the end, one is forced to recognize that there is just one structure; understanding of the physical universe is all of one piece.

PHYSICS AND OTHER SCIENCES

Physics is in many ways the parent of the other physical sciences, but the relation is a continually changing one. Modern chemistry is permeated with ideas that came from physics, so thoroughly permeated, in fact, that the sudden demise of all physicists—with the exception of an important class calling themselves chemical physicists—would not immediately slow down the application of physical theory to chemical problems. The last great theoretical contribution of physics to chemistry was quantum mechanics. For another such contribution there is no room, almost by the definition of chemistry. From the point of view of the physicist, chemistry is the study of complex systems dominated by electrical forces. Strong interactions, weak interactions, gravitation—these are of no direct interest to the chemist. There is no reason to doubt, and voluminous evidence to show, that quantum mechanics and electromagnetic theory as now formulated provide a complete theoretical foundation for the understanding of the interactions between atoms and molecules.

An immense task remains for the theoretical chemist, a task that is in some part shared by the physicist interested in the same problems. One area of common interest is statistical mechanics, especially the theory of "cooperative" phenomena such as condensation and crystallization, where, although the forces that act between adjacent molecules are known, the behavior of the whole assembly presents a theoretical problem of singular subtlety. Other problems that attract both chemists and physicists include phenomena on surfaces, properties of polymers, and the fine details of the structure and spectra of simple molecules. A subject of very intense research in which physics and chemistry are thoroughly blended is the study, both experimental and theoretical, of reactions in rarefied partially ionized gases. This study has direct applications in plasma physics, the development of high-power lasers, the physics of the upper atmosphere, and astrophysics.

There is really no definable boundary between physics and chemistry.

[13]

There never has been. Approximately 5000 American scientists, on a rough estimate, are engaged in research that would not be out of place in either a physics or a chemistry department. Some call themselves physicists and their specialty chemical physics or just physics. Others are physical chemists. The label generally reflects the individual's graduate training and correlates with some differences of interest and style. These chemical physicists have illustrious predecessors, including Michael Faraday and Willard Gibbs. And those who, like them, have made a permanent mark on both sciences are likely to be thought of as physicists by physicists and chemists by chemists.

Physics serves chemistry in quite another way. It is the source of most of the sophisticated instruments that the modern chemist uses. This dependence on physics has been, if anything, increasing. Perhaps the infrared spectrograph and the x-ray diffraction apparatus should be credited to the physics of an earlier era; their present highly refined form is largely the result of commercial development stimulated by users. But mass spectrographs, magnetic resonance equipment, and microwave spectrometers, all of which originated in physics laboratories in relatively recent times, are found in profusion as well as are the more general electronic components for detecting photons and atoms—electron multipliers, low-noise amplifiers, frequency standards, and high-vacuum instrumentation. One might follow a research chemist around all day, from spectrograph to computer to electronic shop to vacuum chamber, without deducing from external evidence that he was not a physicist, unless, as might still happen today, the smell of his environment gave it away.

Radiochemistry is in a class by itself. The radiochemist and the nuclear physicist have been partners indispensable to one another since before either specialty had a name. The dependence of experimental nuclear physics on radiochemical operations is perhaps less conspicuous, seen against the whole enterprise of nuclear physics, than it was 10 or 20 years ago. On the other hand, advances in the use of labeled elements and compounds in chemical, biochemical, and medical research continue to be paced by improvements in detection methods. These came directly from physics. A spectacular recent example is the solid-state particle detector, with parentage in nuclear physics and solid-state physics.

Instead of viewing physics and chemistry as different though related sciences, it might make more sense to consider a science of substances, with its base in quantum physics and objects of study ranging from the crystalline semiconductor (now assigned to the solid-state physicist) to the alloys of the metallurgist, to the molecular chain of high-polymer physics and chemistry, to the elaborate molecular structures of the organic chemist. Through this whole range of inquiry one can discern a remarkable convergence in theoretical treatment, and also in experimental methods. The first comes about as fundamental understanding replaces phenomenology. When the properties of the complex system, be it a boron whisker or a protein molecule, can be systematically deduced from the arrangement of its elementary parts, which are nothing but atoms governed by quantum mechanics, a universal theory of ordinary substances will be at hand. Such a theory has not yet been achieved, but as theoretical methods become more powerful, they become, as a rule, more general, and there is steady progress in that direction. Already the language of theory in organic chemistry is much closer than it used to be to the language of theory in solid-state physics.

The convergence in experimental methods, which of course should never become complete, also reflects the tendency of more powerful analytic methods to be more general. The scanning electron microscope is equally precious to the biochemist and the metallurgist. The infrared spectrograph is almost as ubiquitous as the analytic balance. Radioactive labeling is practiced in nearly all the physical sciences.

Notwithstanding the staggering accumulation of detailed information in the materials sciences, a drastic simplification of scientific knowledge is occurring in these fields. As the facts multiply, the basic principles needed to understand them all are being consolidated. To be sure, the need for specialization by individuals is not declining; the quantity of information vastly exceeds what one mind can assimilate. But the specialist is no longer the custodian of esoteric doctrine and techniques peculiar to his class of substances. Quantum physics is replacing the cookbook, and the mass spectrograph is replacing the nose. The future organic chemist acquires a rough working knowledge of quantum mechanics very early—often earlier, the physicist must concede with chagrin, than his roommate who is majoring in physics. Soon, if it is not already so, any single section of this enormously rich and varied picture will be understood at a fundamental level by anyone equipped with a certain common set of intellectual tools.

Well under way here is nothing less than the unification of the physical sciences. This unification is surely one of the great scientific achievements of our time, seldom recognized or celebrated, perhaps, because, having progressed so gradually, it cannot be seen as an event. Nor can it be credited to one science alone. The influence of quantum physics on chemistry was clearly a central development, and, if one wishes to symbolize that development by one of its landmarks, there is Linus Pauling's *The Nature of the Chemical Bond*. In physics there are many landmarks in the theory of condensed matter, from the first application of quantum theory to crystals by Einstein and Debye to the solution of the riddle of

[15]

superconductivity, among them the quantum theory of metals, the understanding of ferromagnetism, and the discovery of the significance of lattice imperfections in crystals. But the basic contribution of physics is the secure foundation on which all this knowledge is built—on understanding, confirmed by the most stringent experimental tests, of the interactions between elementary particles and the ways in which they determine the structure of atoms and molecules. The fruits of this immense achievement are only beginning to appear.

Biology obviously derives part of its nourishment from physics by way of chemistry. Biochemistry and molecular biology are equally dependent on physical instrumentation. X-ray diffraction, electron microscopy, and isotopic labeling are indispensable tools. Modern electronics is important in physiology, most conspicuously in neurophysiology, where spectacular progress has been made by observing events in single neurons, made accessible by microelectrodes and sophisticated amplifiers. Other examples are described in the Report of the Panel on Physics in Biology.

These are products of past physics. One might wonder whether future physics is likely to prove as fruitful a source of new experimental techniques for biology and medical science. There are two reasons for thinking that it will. First, there is no apparent slackening of the pace of innovation in experimental physics. In almost every observational dimension, short time, small distance, weak signal, and the like, the limits are being pushed beyond what might have been reasonably anticipated. If there is one thing experience teaches here, it is that quite unforeseen applications eventually develop from any major advance in experimental power. Through the Mössbauer effect, preposterous as it seems, motions as slow as that of the hand of a watch can be measured by the Doppler shift of nuclear gamma radiation. Even after this discovery, when Mössbauer experiments were going on in dozens of nuclear-physics laboratories, a physiological application would have seemed rather fanciful. In fact, the Mössbauer effect is being used today to study, in the living animal, the motion of the basilar membrane in the cochlea of the inner ear, perhaps the central problem in the physiology of hearing.

There is another reason to look forward to contributions to the life sciences from inventions not yet made. It is the existence of some obvious and rather general needs, the satisfaction of which would not violate fundamental physical laws, for example, an x-ray microscope with which material could be examined *in vivo* with a resolution of, say, 10 Å or a better way of seeing inside the body than the dim shadowgraphs, remarkably little better than the first efforts of Roentgen, that medical science has had to be content with for half a century. But the breakthroughs probably will again come in unexpected ways; one cannot guess what will

[16]

play the role of Roentgen's Crookes tube. The physicist can only feel rather confident that an active, inventive period in experimental physics eventually will have important effects on the way research is done in the biological sciences.

The intellectual relations between physics and biology are changing, perhaps more because of what is happening in biology than what is happening in physics. Most physicists who have any acquaintance with biology, if only through semipopular accounts of the latest discoveries, find the ideas of current biology, especially molecular biology, intriguing and stimulating. No physicist could fail to be stirred by the elucidation of the genetic code or by the other glimpses into primary mechanisms of life. This wonderful apparatus works by physics and chemistry after all! But it is far more ingenious and subtle than any contrivance of wires, pulleys, and batteries. From the intricate engine of muscle fiber to the marvelous information processer in the eye, plainly there are hundreds of mechanisms in which physics, chemistry, and biological function are inextricably involved. Also, the evident universality of basic processes in the cell appeals strongly to a mind trained in the physicist's approach to structure and function. There is no doubt that biology is going to attract some students who would have made good physicists, which cannot be deplored. It is to be hoped that there soon will be a growing number of biologists who are not only well grounded in physics but who share, and possibly derive some encouragement from, the physicist's conviction that the behavior of matter can be understood in terms of the interactions of its elements; this behavior and these interactions are the goals of experimental study.

At the other end of the scale is astronomy. Physics began with astronomy, but after the foundations of Newtonian mechanics were secured, astronomical observations (not counting as such the observations of cosmic rays) did not directly generate new fundamental physics. However, astronomy did provide a rich field for the application of physics. Great advances in astronomy such as the elucidation of the structure and evolution of stars depended on an understanding of the structure of atoms. That came from the physics laboratory and from quantum theory as it developed. Then it was nuclear physics that supplied the keys to the generation of energy in the stars and to the production of the elements. These questions were highly interesting to physicists and inspired both theoretical and experimental work. But, broadly speaking, this work was merely physics applied to astronomical problems.

At a different level, though, astronomy has always had a powerful intellectual influence on physics. The heavens confronted man with tantalizing mysteries. His conceptions of what he saw there strongly influenced

[17]

philosophical attitudes toward nature. Astronomy has given the physicist confidence that the universe at large is governed by beautifully simple laws of physics, discoverable from earth by man. That belief gives the explorations of physics a wider purpose and significance. It attracts the physicist's attention to cosmological questions, to the physics of gravitation, and to phenomena occurring under conditions utterly unattainable in a terrestrial laboratory.

Today the interaction of physics and astronomy is more vigorous than at any time since Newton. Astronomy has entered an astonishingly rich period of significant discovery. This is due, in part, to observing over a greatly widened spectrum, from the long waves of radio astronomy, which have in 25 years greatly increased the knowledge of the large-scale universe, to x rays and gamma rays, which are just beginning to produce interesting information. In part, too, it reflects the increased power and scope of astrophysical theory, working from a more complete base in atomic and nuclear physics. Also, nature has provided some incredibly marvelous, totally unexpected features for telescopes to discover, displaying on a grand scale phenomena that involve most of physics. Less than ten years after the maser was invented in a physics laboratory, the maser process was found to occur in clouds of interstellar gas. It is typical of the present intensive involvement of physicists in astronomy that this discovery was made by some of the same physicists, now turned radio astronomers, who had participated in the microwave spectroscopy that led to the invention of the maser.

Nuclear physicists and astrophysicists have been engaged for more than 20 years in a collaboration from which has come not only an understanding of the source of energy in stars but of the production of the chemical elements found in the universe. This knowledge bears directly on the history of the universe, providing much of the solid evidence against which cosmological theories can be tested. More surprising is the emerging importance to astronomy of elementary-particle physics. The opacity of matter to neutrinos turns out to be relevant not only to the reconstruction of a primordial big bang but to what is going on now at the centers of galaxies. Inside pulsars there is almost certainly "hyperonic" matter, composed of particles more massive than protons, known only in the laboratory as evanescent products of high-energy collisions. Perhaps a not negligible fraction of the matter in the universe is compressed into this state, a form of matter hardly speculated about before pulsars were discovered five years ago. It may be difficult to forecast commercial applications on planet Earth for high-energy physics, but its importance in the universe as a whole may have been greatly underestimated.

Of course, cosmic rays have been studied by physicists, not astronomers,

[18]

for 50 years; and these particles, still the most energetic a physicist can hope to see, have been transcendentally important in the development of modern physics. It is hard to imagine how elementary-particle physics would have progressed if the earth had been shielded from cosmic rays. Although the source of cosmic rays was obviously astronomical, it is only rather recently that the importance of cosmic radiation as a constituent of the interstellar medium has been appreciated. Something like a merger of cosmic-ray and related high-energy physics with astrophysics has taken place; the new Division of Cosmic Physics in the American Physical Society is one indication. Magnetohydrodynamics and plasma physics are very lively subjects of common interest to members of both groups. Beyond these obvious cases of interest, even solid-state physicists have been drawn into astrophysics by the discovery of neutron stars.

In the same period, a resurgence of interest in gravitation has occurred among both astronomers and physicists—among astronomers because of the discovery of systems close to the theoretical conditions for gravitational collapse and among physicists because of experimental developments that bring some predictions of gravitational theory within the range of significant laboratory test.

All these developments are bringing again to physics and astronomy a wonderful unity of interest. Never before have so many parts of physics directly concerned astrophysicists; seldom before have astronomical phenomena so stirred the imagination of physicists.

The cosmos is still the place where man must look for answers to some of the deepest questions of physics. Were the fundamental ratios that characterize the structure of matter as found here and now truly precisely constant for all time? Observations of distant galaxies offer a view backward in time to an earlier stage of the universe. Is Einstein's general relativity an exact and complete description of gravitation? Is the visible universe a mixture of matter and antimatter in equal parts, or is what is called matter overwhelmingly more abundant throughout? Already astronomers have observations that bear on these questions. The conclusions are only tentative now, but it seems quite certain that the questions will be answered.

As for the earth sciences, a gap no longer exists between astronomy and geology. A look at the relation of physics to the earth sciences shows a network of interconnected problems, stretching from the center of the earth to the center of the galaxy. The earth's magnetic field provides a good example. How it is generated has always been a puzzle. Now it appears, although the explanation is not complete, that magnetohydrodynamic theory is about to produce a convincing picture of the electric dynamo that must be at work within the earth's fluid core. Furthermore, the same ideas may explain the generation of magnetic fields of stars and even, when applied on a very different scale, the magnetic fields that pervade the whole galaxy. These developments are the work of both geophysicists and astrophysicists, many of them people whose breadth of interest would justify both titles. In addition, the interplanetary magnetic fields in the solar system, which are dominated by the solar wind, are of interest to both the planetary physicist and the solar physicist.

The theoretical base for these interrelationships is the dynamics of highly conducting fluids, including ionized gases, which is also the base for such potentially important engineering developments as the magnetohydrodynamic generator. Not new fundamental physics but ingenious and insightful analysis and the development of more powerful theoretical tools are needed.

From the point of view of physics, the other sciences might be grouped into four very broad divisions: a science of substances, including chemistry and also a part of physics; life sciences; earth sciences plus astronomy (for which a good name that will comprehend the range from meteorology to cosmology is lacking); and engineering science. These divisions are, of course, multiply overlapping, with a topology that would defy a two-dimensional diagram. A category of current interest, environmental science, would overlap all four.

Engineering science is suggested as a fourth division, although it is not as extensive or as well recognized as the others, to emphasize a distinction between the products of technology and the growing body of knowledge— scientific knowledge—that constitutes the intellectual capital of engineering. To this knowledge both engineers and physicists contribute continually, with a mutual stimulation of ideas. To call different portions of this body of knowledge mere applied mathematics does not do justice to the imaginative work that goes on or to the potential influence on the other sciences of the ideas generated. A previously mentioned example is the important subject of fluid dynamics, with its ubiquitous problem of turbulence, in which engineering science naturally has a big stake. Consider, as another example, communication theory, developed in its many aspects by people calling themselves variously engineers, mathematicians, and physicists. Sophisticated treatments of fluctuation phenomena, including quantum effects, the relation of information to entropy, and the rich ramifications— including holography—of Fourier duality are just a few of the ideas it encompasses. Or consider the theory of automatic control with feedback, which was developed mainly within engineering science but is now an indispensable aid in most experimental sciences, including physics. The point is that between physics and engineering science there are strong intellectual, one might even say cultural, links. That is only one aspect of the relation

[20]

of physics to technology, a topic explored more fully in the following section.

No one would question the importance of physics in the development of these fields of science. However, because chemistry needed physics does not necessarily imply that chemistry now needs help from physicists. Physicists have made fairly direct contributions to chemistry, even recently; and physics, at least as a source of new experimental tools and techniques for other sciences, may be as fruitful a source in the immediate future as it has been in the past. But can essential contributions from physics to the other sciences in the form of new and basic ideas be expected? Do the chemists or the earth scientists, who have fairly well assimilated the apparently relevant parts of physics, need the physicist for any service except to teach physics to their students? What is their interest in his hunt for quarks or gravity waves?

There are two ways to answer such questions from the physicist's point of view. One can meet example with example, explaining, for instance (as will be done in Chapter 4 when this question is addressed with specific reference to high-energy physics), how the isolation of the quark could have immense practical consequences. Or one can make a more general reply along the following lines. The increasing unity of the physical sciences at the basic level and the proliferation of interconnections among the fields, and especially with physics, make intellectual vigor widely contagious. New ideas tend to stimulate other new ideas. As long as physics has great questions to work on, its discoveries can hardly fail to excite resonances in neighboring fields.

TECHNOLOGY AND PHYSICS: THEIR MUTUAL DEPENDENCE

Everyone knows that today the main sources of new technology are research laboratories of physics and chemistry and not the legendary ingenious mechanic or Edisonian wizard. Actually, the relation of technology, that is, applied science, to basic science has been close for more than a century. Think of Faraday, Kelvin, Pasteur. It is true that Morse and Bell were amateurs in electricity, while Maxwell, it is said, found the newly invented telephone not interesting enough to serve as a subject of a scientific lecture. But the sweeping exploitation of electromagnetism that began in the latter half of the nineteenth century was based directly on the fundamental understanding achieved by Maxwell. No one would suggest that today's semiconductor technology could have been created solely by engineers ignorant of the relevant fundamental physics. Research in physics provided the base from which present technology is developing.

[21]

But physics research did, and is doing, more than that. Research is a powerful stimulator of fresh ideas. One reason is that in research, and especially in the most fundamental research, the scientist is often trying to break new ground. He may need to measure something at higher energy (remember Van de Graaff and the electrostatic generator) or closer to absolute zero (Kamerlingh Onnes discovering superconductivity) or in a previously inaccessible band of the spectrum. Years before World War II the magnetron was first exploited for the generation of 1-cm waves by the physicists Cleeton and Williams at the University of Michigan. They used it to make the first observation of the inversion resonance of the ammonia molecule.

Also, and this applies to both experimental and theoretical research, to be challenged by a puzzling phenomenon stimulates the imagination. One is likely to try looking at things from a new angle, questioning assumptions that had been taken for granted. P. W. Bridgman once described the scientific method as "the use of the mind with no holds barred." The uninhibited approach of the research scientist to a strange problem has even generated a whole discipline—operations research. Prominent among its creators were scientists like P. M. S. Blackett and E. G. Williams, who came from fundamental physics research, both experimental and theoretical, of the purest strain.

The research laboratory, including the theoretical physicist's blackboard or lunch table, provides the kind of freewheeling environment in which an idea can be followed for a time to see where it leads. Most new ideas are not good. In a lively research group these are quickly exposed and discarded, often having stimulated a fresh idea that may be more productive.

In such a setting, physicists are not generally intellectually constrained by the distinction between fundamental science and technology. For one thing, experimental physics heavily depends on some very advanced technology. The research physicist is not only at home with it, he has often helped to develop it, adapt it, and debug it. He is part engineer by necessity—and often by taste as well. An experimental physicist who is totally unmoved by a piece of excellent engineering has probably chosen the wrong career. One cannot make such a sweeping statement about theoretical physicists, but even they, as was spectacularly demonstrated long ago in the Manhattan Project, frequently can apply themselves both effectively and zestfully to technological problems. Currently, in fields such as plasma physics and thermonuclear research, there are many theoretical physicists, with a broad range of interest and expertise, some with a background in elementary-particle physics, intimately concerned with engineering questions.

Ongoing basic research is necessary for the translation of scientific discovery into useful technology, even after the discovery has been made. As a rule, the eventual value to technology of a discovery is seldom clearly evident at the time. It often emerges only after a considerable evolution within the context of fundamental research, sometimes as an unexpected by-product. Nuclear magnetic resonance (NMR) is now widely used in the chemical industry for molecular structure identification. This possibility was totally unforeseeable in the early years of NMR research. It came to light only after a major improvement in resolution had been achieved by physicists studying NMR for quite different purposes. However, a backlog of unapplied basic physics is not all that it takes to generate new technology; it may not even be the main ingredient.

Some of the most startling technological advances in our time are closely associated with basic research. As compared with 25 years ago, the highest vacuum readily achievable has improved more than a thousandfold; materials can be manufactured that are 100 times purer; the submicroscopic world can be seen at 10 times higher magnification; the detection of trace impurities is hundreds of times more sensitive; the identification of molecular species (as in various forms of chromatography) is immeasurably advanced. These examples are only a small sample. All these developments have occurred since the introduction of the automatic transmission in automotive engineering!

On the other hand, fundamental research in physics is crucially dependent on advanced technology, and is becoming more so. Historical examples are overwhelmingly numerous. The postwar resurgence in low-temperature physics depended on the commercial production of the Collins liquefier, a technological achievement that also helped to launch an era of cryogenic engineering. And today, superconducting magnets for a giant bubble chamber are available only because of the strenuous industrial effort that followed the discovery of hard superconductors. In experimental nuclear physics, high-energy physics, and astronomy—in fact, wherever photons are counted, which includes much of fundamental physics—photomultiplier technology has often paced experimental progress. The multidirectional impact of semiconductor technology on experimental physics is obvious. In several branches of fundamental physics it extends from the particle detector through nanosecond circuitry to the computer output of analyzed data. Most critical experiments planned today, if they had to be constrained within the technology of even ten years ago, would be seriously compromised.

The symbiotic relation of physics and technology involves much more than the exchange of goods in the shape of advanced instruments traded for basic ideas. They share an atmosphere the invigorating quality of which depends on the liveliness of both. The mutual stimulation is most

[23]

obvious in the large industrial laboratory in which new technology and new physics often come from the same building and, sometimes, from the same heads. In fact, physics and the most advanced technology are so closely coupled, as observed, for example, on a five- to ten-year time scale, that the sustained productivity of one is critically dependent on the vigor of the other.

EXPERIMENTAL PHYSICS

An experimental physicist is usually doing something that has not been done before or is preparing to do it, which may take longer than the actual doing. That is not to say that every worthwhile experiment is a risky venture into the unknown. Many fairly straightforward measurements have to be made. But only *fairly* straightforward! The easy and obvious, whether in basic or applied physics, has usually been done. The research physicist is continually being challenged by experimental problems to which no handbook provides a guide. Very often he is trying to extend the range of observation and measurement beyond previous experience.

A most spectacular example is the steady increase in energy of accelerated particles from the 200-keV protons, with which Cockcroft and Walton produced the first artificial nuclear disintegrations in 1930, to the 200-GeV protons of the National Accelerator Laboratory—in 40 years a factor of a million! This stupendous advance was achieved not in many small steps but in many large steps, each made possible by remarkable inventions and bold engineering innovations produced by physicists. Although numerical factors of increase do not have the same implications in different technologies, few branches of engineering have come close to that record. A possible exception is communication engineering. Transmission by modulated visible light, now feasible, represents an increase in carrier frequency of roughly a million over the highest radio frequency usable 40 years ago. What is really more significant, the information bandwidth achievable has increased by a comparable factor. This great advance was, of course, made possible by development in basic physics and at many stages was directly stimulated by the basic research of physicists. Even accelerator physics contributed at one stage by stimulating the development of klystrons. The accelerator physicists have a remarkable record of practical success as engineers. From the time of the early cyclotrons to the present, no major accelerator in this country, however novel, has failed to work; most of them exceeded their promised performance.

In other branches of experimental physics, too, people are doing things that would have appeared ridiculously impractical only 10 to 20 years ago. For example, a discovery in the physics of superconductivity (the Josephson effect) made it possible to measure precisely electric voltages and magnetic fields a thousand times weaker than could be measured previously. Electrons, also positive ions, can be electrically caged, almost at rest in space, for hours. By another technique, neutral atoms can be stored in an evacuated box for minutes without disturbing a natural internal oscillator that completes, in that time, about 10^{12} cycles of oscillation. One by-product is an atomic clock that is accurate to about a second in a million years. By recent laser techniques, light pulses of only 10^{-12} seconds' duration have been generated and observed. The local intensity of light that can be created with lasers is many million times greater than anything known in the laboratory ten years ago. Effects can be readily observed that in the past could be only the subjects of theoretical speculation, thus opening to investigation as entire field of basic research—nonlinear optics. Within the same decade, the highest magnetic-field strengths easily available in the laboratory, which had hardly changed in a century, were roughly quadrupled by superconducting magnets. In the same period, low-temperature physicists extended downward by a factor of 10 the temperature range usable for general experimentation.

These developments and others mentioned subsequently in this Report show that experimental physics is not running out of ideas or becoming a routine matter of data-gathering. In fact, experimental physics could be entering a new period, distinguishable (by criteria other than austerity of budgets!) from two preceding periods of conspicuous experimental advance in modern physics: the decade before World War II, which, in terms of tools and techniques, could be called the "cyclotron and vacuum tube" period, and the immediate postwar period, in which microwave electronics and nuclear technology, largely the fruits of wartime physics, made possible an enormous advance in experimental range. Within the past 10 to 15 years, several postwar developments have come of age that, considered as a group, promise a comparable advance in experimental capability. These include cryogenics in all its ramifications, semiconductor technology, laser–maser techniques, and the massive exploitation of computers. One trouble with any such historical formula, and the glory of physics as an adventure, is the existence of the important and unclassifiable exceptions. For example, the continuously spectacular progress in high-energy accelerators fits only very loosely into the scheme just outlined. A more modest exception is the simple proportional counter, one of the most elegant and sensitive devices of physics since the torsion pendulum, which has survived through all three periods, earning in each a new lease on life.

Whether it signifies a new era or not, the enormous advance in observational power that is occurring now will in all likelihood open still more fields of research in physics. It will certainly lead to applications yet unforeseen in other sciences and technology.

PHYSICS—A CONTINUING CHALLENGE

It is possible to think of fundamental physics as eventually becoming complete. There is only one universe to investigate, and physics, unlike mathematics, cannot be indefinitely spun out purely by inventions of the mind. The logical relation of physics to chemistry and the other sciences it underlies is such that physics should be the first chapter to be completed. No one can say exactly what completed should mean in that context, which may be sufficient evidence that the end is at least not imminent. But some sequence such as the following might be vaguely imagined: The nature of the elementary particles becomes known in self-evident totality, turning out by its very structure to preclude the existence of hidden features. Meanwhile, gravitation becomes well understood and its relation to the stronger forces elucidated. No mysteries remain in the hierarchy of forces, which stands revealed as the different aspects of one logically consistent pattern. In that imagined ideal state of knowledge, no conceivable experiment could give a surprising result. At least no experiment could that tested only fundamental physical laws. Some unsolved problems might remain in the domain earlier characterized as organized complexity, but these would become the responsibility of the biophysicist or the astrophysicist. Basic physics would be complete; not only that, it would be manifestly complete, rather like the present state of Euclidean geometry. Such an outcome might not be logically possible.

One might be more seriously concerned with the prospect of reaching a stage short of that, in which all the basic physics has been learned that is needed to predict the behavior of matter under all the conditions scientists find in the universe or have any reason to create. From chemistry to cosmology, let us suppose, all situations are covered, but one cannot predict with certainty the scattering cross section for e-neutrinos on μ-neutrinos at 10^{30} V. Suppose further that the experiments required to explore fully all the physics at 10^{30} V are inordinately costly and offer no prospect of significantly improving physics below 10^{20} V, which is already known to be sufficiently reliable. If some such state were reached, one might reasonably expect that research in fundamental physics would be at least brought to an indefinite halt if not closed out entirely as being in the state of perfection previously postulated. It would be said that all the physics that mattered had been learned.

[26]

Some physicists are supposed to have made this statement about physics at about the end of the nineteenth century. That they were wrong, spectacularly wrong, is a reminder that human vision is limited; it proves nothing more. For the state of knowledge of physics today is essentially different from that in 1890, just before the curtain was pulled back, so to speak, from the atomic world. Where the problems lie is evident. As far as the behavior of ordinary matter is concerned, it is hardly conceivable that the detailed picture of atomic structure, the product of quantum theory and exhaustive experimentation, should turn out to be misleading or that the main problem in nuclear physics should suddenly be revealed as one hitherto ignored. There are mysteries, but they lie deeper. If it were possible to fence off a particular range of application, for example, chemistry at temperatures below 10^5 deg, then a state in which all the relevant fundamental physics is essentially complete could be reasonably anticipated. Indeed, for a sufficiently restricted application, the day might already have arrived.

The trouble is that the range of interests continues to widen, and in unexpected ways. Because of pulsars, the structure of atoms exposed to a magnetic field of 10^{12} G (ten million times the strongest fields in laboratory magnets) becomes a question of some practical concern, as does the shear strength of iron squeezed to a billion times its ordinary density. Just now astronomy seems to be making the most new demands on fundamental physics; there the end is not in sight.

Even if the physicist could reliably and accurately describe any elementary interaction in which a chemist or an astronomer might be interested, the task of physics would not be finished. Man's curiosity would not be satisfied. Some of the most profound questions physics has faced would remain to be answered, if understanding of the pattern of order found in the universe is ever to be achieved. The extent of present ignorance still is great.

It is far from certain that in the presently recognized elementary particles the ultimate universal building blocks of matter have been identified. The laws that govern the behavior of the known particles under all circumstances are not known. It is even conceivable that the study of particle interactions at ever higher energy leads into an open domain of never-decreasing complexity. Probably most physicists would doubt that. Cosmic rays afford an occasional glimpse of matter interacting at energies very much greater than particle accelerators provide, and no bizarre consequences have yet been observed. It seems rather that physicists now face not mere complexity but subtlety, a strangeness of relationship among the identified particles that might render the question of which of them is truly elementary essentially meaningless.

[27]

Even if physicists could be sure that they had identified all the particles that can exist, some obviously fundamental questions would remain. Why, for instance, does a certain universal ratio in atomic physics have the particular value 137.036 and not some other value? This is an experimental result; the precision of the experiments extends today to these six figures. Among other things, this number relates the extent or size of the electron to the size of the atom, and that in turn to the wavelength of light emitted. From astronomical observation it is known that this fundamental ratio has the same numerical value for atoms a billion years away in space and time. As yet there is no reason to doubt that other fundamental ratios, such as the ratio of the mass of the proton to that of the electron, are as uniform throughout the universe as is the geometrical ratio $\pi = 3.14159$. Could it be that such physical ratios are really, like π, mathematical aspects of some underlying logical structure? If so, physicists are not much better off than people who must resort to wrapping a string around a cylinder to determine the value of π! For theoretical physics thus far sheds hardly a glimmer of light on this question.

The question was posed in even sharper form 40 years ago by Eddington, who argued that the structure perceived in nature can be nothing but a reflection of the methods of observation and description that must be employed. That view would reduce fundamental physics to metaphysics. But Eddington's own conception of the structure did not survive. Such evidence as he had adduced was soon washed away in a flood of discovery. The whole history of physics since then gives no sign that physics is about to become an exercise in deduction. Every attempt to close the theoretical structure to all changes except refinements has been confounded by an experimental discovery. This has happened so often that there has been some accession of intellectual humility along with the vast increase in knowledge of the underlying structure of matter. Surely the end of the story is yet far off.

The fundamental question survives, if not the attempts to answer it: Is there an irreducible base, or design, from which all physics logically follows? The history of modern physics warns that the answer to such a question will not be attained just by thinking about it. To be sure, brilliant theoretical ideas, probably many, will be needed, and some future Bohr or Einstein may become renowned for the flash of insight that eventually reveals a key to the puzzle (or the absence of a puzzle!). But without experimental exploration and discovery, new ideas are not generated. Physics will remain an experimental science at least until very much more is known about the fundamental nature of matter.

[28]

4

THE

SUBFIELDS

OF

PHYSICS

INTRODUCTION

This chapter provides a brief status report on physics and its subfields, including the historical background that has brought each subfield to its present status, current scientific activity in each subfield, and the distribution of activity in each.

Although each subfield is considered in some detail, to bring out its internal logic, emphasis is placed on the underlying unity of physics. Much has been said in recent years about the extent to which physics and, indeed, many of the other sciences were fragmenting with ever-increasing specialization, so that effective communication within and among sciences is rapidly dwindling. We believe that, although there are obvious dangers that must be guarded against, the often discussed fragmentation of physics is exaggerated. For the entire field of physics is tightly linked by common techniques—both experimental and theoretical—but most of all by a common style or approach to problems. With increasing specialization, what tends to be forgotten is the remarkable extent to which techniques and concepts diffuse rapidly throughout physics and the degree of commonality that exists. This underlying unity is illustrated in the closing section of this chapter in discussions of the four spectroscopies and of one of the most exciting new objects found in nature—the pulsar.

The basic principles of classical and quantum mechanics and of relativity provide a unifying framework for all activity in physics. Modern physics encompasses a remarkable range in both space and time, as illustrated in Figure 4.1. The characteristic dimensions of the entities of physics span a range of some 10^{40}; strangely enough, the range of characteristic times—from the lifetime of the metagalaxy to a time characteristic of subnuclear phenomena, say the passage of a photon over a typical elementary particle—is again some 10^{40}. It is these enormous ranges that provide physics with much of its richness and challenge.

In moving from the macroscopic systems of classical physics and astronomy to the atomic domain, physicists found no necessity to add to their two natural forces—gravitation and electromagnetism. But they were

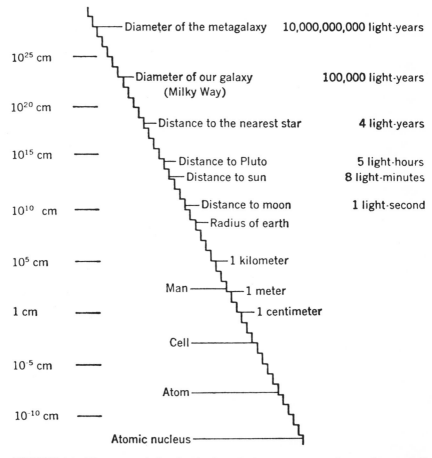

FIGURE 4.1   The range of sizes in physics. Each step corresponds to a factor of 10 increase in size.

[30]

forced to develop a new arithmetic—quantum mechanics—whose assumptions reflect the indistinguishability of the microscopic constituents of matter and the abandonment of mechanical determinism. On moving deeper into the microcosm to the nuclear realm, physicists found that they had to double the number of kinds of natural forces, adding the weak and the strong nuclear interactions, but they were pleased to find that quantum mechanics, augmented with appropriate relativistic corrections required for high accuracy, was entirely adequate to describe the nuclear system, just as it had been in the case of the atom.

In moving on to elementary-particle physics, relativity has come to the forefront in all considerations. Fundamental interest has centered on whether additional natural forces, or modifications of quantum mechanics, or, more generally, quantum electrodynamics (QED) would be required to reproduce the new phenomena. Much effort has been devoted to these questions. At present there are hints that a superweak nuclear force may be required, but there is no evidence yet to suggest that QED is not entirely appropriate for the description of all phenomena in the range of energies and linear dimensions thus far accessible.

Ignoring for the moment the possible existence of a superweak nuclear force whose effects would be entirely negligible in all save the most esoteric situations, physicists have a mathematical framework—quantum electrodynamics—and four natural forces—gravitation, electromagnetism, and the two nuclear forces, strong and weak—within which they attempt to encompass and explain all natural phenomena (Plate 4.I). This economy of basic input and this focus on fundamental phenomena are the hallmarks of physics. But no less a part of physics is the development and application of new insights; in both areas, fundamental and applied, physics has played and continues to play an important role in education, in interaction with other sciences, in interaction with technology, and in addressing pressing problems facing man and society.

In developing this Report the Survey Committee early decided that it was essential to obtain detailed input from panels of experts in each of a number of subfields that it had identified. Several of these subfields have relatively well-defined and traditional boundaries within physics, for example, acoustics; optics; condensed-matter physics; plasma and fluid physics; atomic, molecular, and electron physics; nuclear physics; and elementary-particle physics. Even here, however, there is a considerable overlap of specific activities within subfields. For example, laser techniques appear in several of the subfields, as do those of colliding atomic and molecular beams. This situation reflects the close coupling within physics, as is emphasized in the section on The Unity of Physics at the end of this chapter. In addition to the subfields noted above, several important interfaces between physics and other sciences were identified.

[31]

There are several areas of classical physics, such as mechanics, heat, thermodynamics, and some elements of statistical physics, that have not been considered explicitly in this Survey. This omission in no way implies any lack of importance of these subjects but merely reflects the fact that they are mature areas of science and that relatively little research *per se* is currently taking place in them.

Early in the Survey, we, as a Committee, developed a lengthy charge, which was addressed to each of the panels. It is included as Appendix B to this volume of the Report. This charge was a broadly ranging one dealing with the structure and activity of each subfield, as viewed not only internally but also in terms of its past, present, and potential contributions to other physics subfields, other sciences, technology, and society generally.

Clearly the charge was most directly relevant to the traditional subfields; in the case of the interface panels some questions were inevitably unanswerable without a survey of equivalent scope of all fields working at the interface. In the case of astronomy, such a survey was available. The panels on data, education, and instrumentation clearly were special cases. As Volume II will show, the panels have responded in depth to the questions asked.

In Appendix 4.A to this chapter, activity in the subfields has been divided into program elements—components large enough to have some internal coherence and reasonable boundaries, and for which it might be possible to estimate present funding levels and PhD manpower involved. As discussed in Chapter 5, the purpose of this exercise was to divide the subfields into units of activity that the Committee could evaluate in terms of intrinsic, extrinsic, and structural criteria. In developing these program elements, the Committee worked with the panel chairmen; however, in some cases the elements used here are not identical with those suggested by the panel chairmen. In the subfields of elementary-particle physics and nuclear physics, it was possible to assign funding levels and manpower rather precisely. Similar assignments for some of the program elements in the other subfields may be in error by a factor of 2.

The principal objective in this chapter is to provide a summary of each of the panel reports, thus giving an overview of U.S. physics—its history, status, opportunities, and problems—as seen by a representative group of active physicists in 1970–1971.

General Activity in Physics

Although manpower, publication, and support data are presented in detail in Chapters 12, 13, and 5 and 10, respectively, to place the discussions of this chapter in perspective we include a number of figures showing some of these data for each subfield (with the exception of interfaces for which data are not fully available). (See Figures 4.2 through 4.8.)

[32]

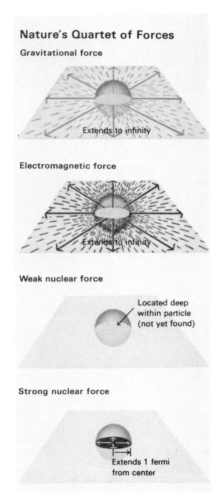

Nature's Quartet of Forces

Gravitational force

Extends to infinity

Electromagnetic force

Extends to infinity

Weak nuclear force

Located deep within particle (not yet found)

Strong nuclear force

Extends 1 fermi from center

PLATE 4.I   Nature's quartet of forces. We can detect a proton only through its forces. It is the sum of four effects. The gravitational force, surrounding all matter in all directions to infinity, controls the stars and galaxies. The much stronger electromagnetic force cancels out at long range, since there are equal numbers of positive and negative charges in the universe. It controls the world of atoms and molecules. The weak nuclear force is known to exist, but its carrier has not yet been detected. The strongest force—the strong nuclear force—controls most effects in the compacted nuclear and subnuclear world. [Source: *Science Year. The World Book Science Annual.* Copyright © 1968, Field Enterprises Educational Corporation.]

[33]

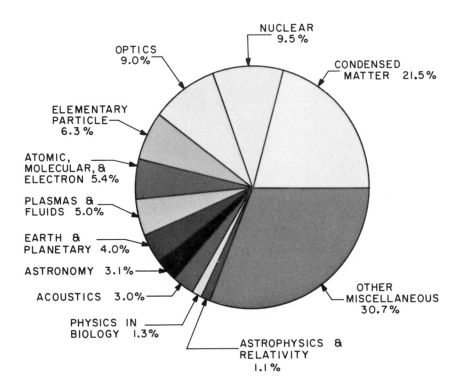

DISTRIBUTION OF PHYSICS MANPOWER
BY SUBFIELD

PhD's and non-PhD's
N = 36,336

NUCLEAR 9.5%

CONDENSED MATTER 21.5%

OPTICS 9.0%

ELEMENTARY PARTICLE 6.3%

ATOMIC, MOLECULAR, & ELECTRON 5.4%

PLASMAS & FLUIDS 5.0%

EARTH & PLANETARY 4.0%

ASTRONOMY 3.1%

ACOUSTICS 3.0%

PHYSICS IN BIOLOGY 1.3%

OTHER MISCELLANEOUS 30.7%

ASTROPHYSICS & RELATIVITY 1.1%

FIGURE 4.2   Distribution of physics manpower (PhD's and non-PhD's) by subfield.
[Source: National Register of Scientific and Technical Personnel, 1970.]

[35]

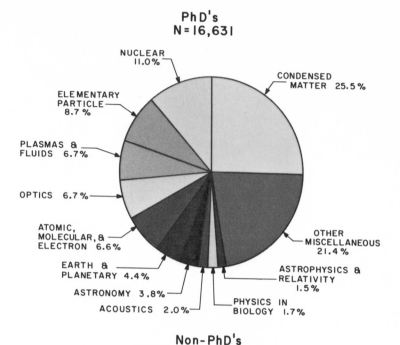

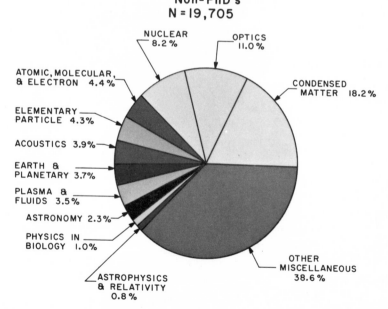

FIGURE 4.3 Distribution of physics PhD's and non-PhD's by subfield. [Source: National Register of Scientific and Technical Personnel, 1970.]

[36]

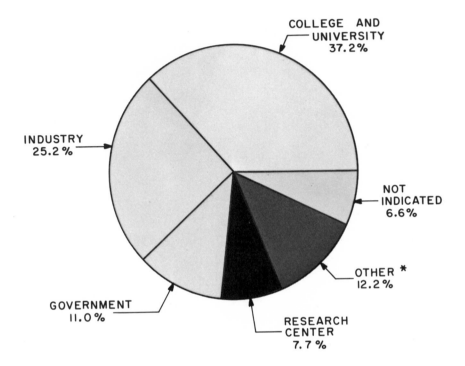

DISTRIBUTION OF PHYSICISTS BY EMPLOYING INSTITUTION

N = 36,336

* Other includes medical school, secondary or elementary school, military service, self-employed, hospital, etc.

FIGURE 4.4 Distribution of physics manpower (PhD's and non-PhD's) by employing institution. [Source: National Register of Scientific and Technical Personnel, 1970.]

[37]

DISTRIBUTION OF PhD AND NON-PhD PHYSICISTS BY EMPLOYING INSTITUTION

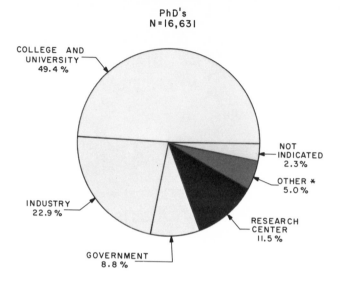

PhD's
N = 16,631

COLLEGE AND
UNIVERSITY
49.4%

NOT
INDICATED
2.3%

OTHER *
5.0%

RESEARCH
CENTER
11.5%

INDUSTRY
22.9%

GOVERNMENT
8.8%

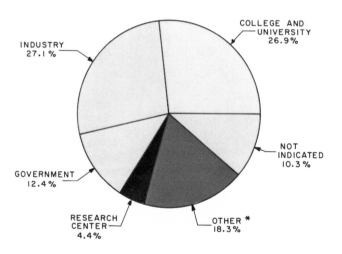

INDUSTRY
27.1%

COLLEGE AND
UNIVERSITY
26.9%

NOT
INDICATED
10.3%

GOVERNMENT
12.4%

RESEARCH
CENTER
4.4%

OTHER *
18.3%

* Other includes medical school, secondary or elementary school,
military service, self-employed, hospital, etc.

FIGURE 4.5   Distribution of physics PhD's and non-PhD's by employing institution.
[Source: National Register of Scientific and Technical Personnel, 1970.]

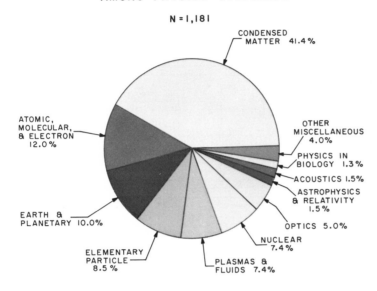

DISTRIBUTION OF PUBLICATIONS
AMONG PHYSICS SUBFIELDS *

N = 1,181

CONDENSED MATTER 41.4%

OTHER MISCELLANEOUS 4.0%

PHYSICS IN BIOLOGY 1.3%

ACOUSTICS 1.5%

ASTROPHYSICS & RELATIVITY 1.5%

OPTICS 5.0%

NUCLEAR 7.4%

PLASMAS & FLUIDS 7.4%

ELEMENTARY PARTICLE 8.5%

EARTH & PLANETARY 10.0%

ATOMIC, MOLECULAR, & ELECTRON 12.0%

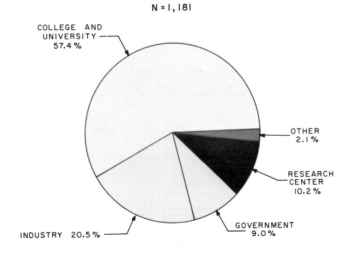

DISTRIBUTION OF PHYSICS PUBLICATIONS
BY ISSUING INSTITUTION *

N = 1,181

COLLEGE AND UNIVERSITY 57.4%

OTHER 2.1%

RESEARCH CENTER 10.2%

GOVERNMENT 9.0%

INDUSTRY 20.5%

* Data are based on a sample of every tenth article listed in 1969 issues of Physics Abstracts.

FIGURE 4.6   Distribution of publications in physics by subfield and by issuing institution.

[39]

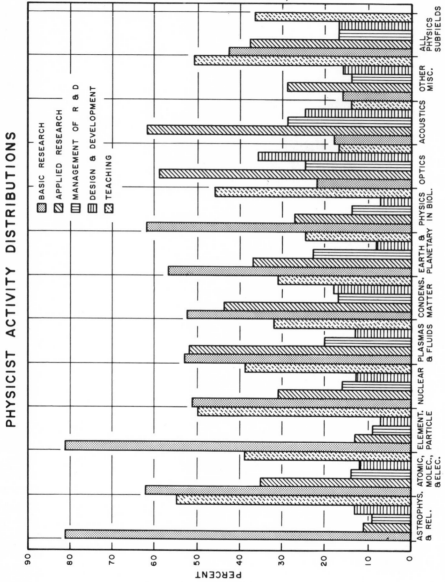

FIGURE 4.7 Principal work activities (those ranked first and second) of physicists (PhD's and non-PhD's taken together) in each subfield. [Source: National Register of Scientific and Technical Personnel, 1970.]

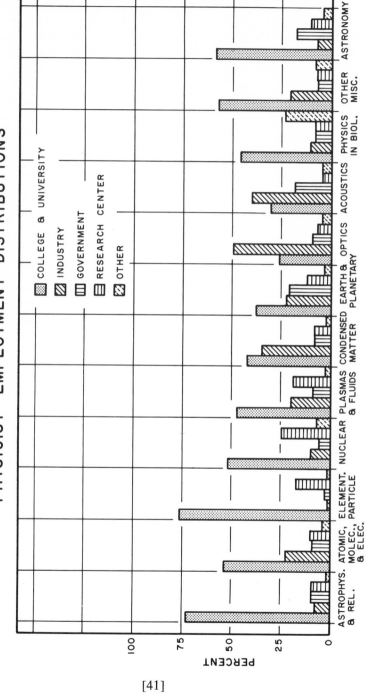

FIGURE 4.8 Distribution of PhD physicists in each subfield by employing institution. [Source: National Register of Scientific and Technical Personnel, 1970.]

The manpower data were derived from the National Register of Scientific and Technical Personnel for 1968 and 1970; the publication data were based on a sample of every tenth article listed in the 1969 issues of *Physics Abstracts;* and the information on support shown here was based largely on the reports of the panels. It should be noted that the totals here need not agree with those appearing in the National Science Foundation's publication, *Federal Funds for Research, Development, and Other Scientific Activities* (FFRDS) and elsewhere in this Report. Some of the panels carried out a more detailed examination of the overall funding pattern, and differences in categories and definitions in some cases have yielded numbers different from those in FFRDS.

Migration data for PhD physicists are displayed in Appendix 4.B to this chapter. As shown there, physics PhD manpower is much more mobile than has been commonly believed. During the period 1968–1970 about one third of the PhD's changed their subfields of major interest. This is a strong indicator of the unity of physics; the subfield interfaces are highly permeable.

> For we do not think that we know a thing until we are acquainted with its primary conditions or first principles and have carried our analysis as far as its simplest elements.
>
> ARISTOTLE (384–322 B.C.)
> *Physics,* Book I, 184

ELEMENTARY-PARTICLE PHYSICS

Introduction

Elementary-particle physics is concerned with the determination of the fundamental constituents of matter and energy, the behavior of matter under the most extreme conditions, the mathematical laws of nature governing this behavior, and the related underlying nature of microscopic space-time itself. The forms of elemental matter that occur naturally, or can be produced in the laboratory, include electrons, photons, protons, and neutrinos. They also include a great (and surprising) variety of unstable particles that are produced in very energetic collisions between the stable particles or between atomic nuclei.

Experimental observations range from measurements of fundamental properties of the particles, made with the greatest imaginable precision, to gross observations of the qualitative behavior or a particular form of matter under the extreme conditions of high-energy impact. Theoretical work includes interpretation of experimental observations based on generally accepted principles, suggestions for new directions of experimentation to answer well-defined questions, the invention of theoretical models and experiments to test them, speculations on the ultimate nature of matter and energy, speculations concerning the generalization of the laws of nature needed to overcome the inadequacies and paradoxes in the existing theory, suggestions of experiments to test these speculations, and deep mathematical investigations of the nature of the theory to reveal its strengths and weaknesses.

Theoretical research in this subfield requires a knowledge of the foundations and methods of most branches of physical theory, including both classical theory and relativistic quantum theory. The required mathematical techniques encompass a wide variety of fields and methods of applied mathematics. Experimental work requires access to a source of particle beams, usually an accelerator, although some types of investigations are carried out successfully with cosmic rays. The experimenter must also have access to very sophisticated instrumentation capable of identifying particles moving at speeds close to that of light or of measuring specific properties of the particles with great precision or of both. The large amount of information that must be sorted to make sense of the behavior of these particles under the extreme conditions to which they are subjected requires the use of very large computers. The development of accelerators (see Figures 4.9 and 4.10), detectors, and other equipment involves a major engineering effort. The people involved in the work include a large coterie of engineers, technicians, programmers, and scanners as well as physicists.

Historical Background

Elementary-particle physics has its roots in all aspects of physics that are historically concerned with the structure of matter, because the questions of structure and constituents of structure are closely related. The interpretation of the term elementary particle has a history going back to the original atomic theory. One need note only the use of the term chemical element, identifying an atomic species, to recognize the beginning of the subject.

The concept of basic building blocks from which the real world can be composed dates back to the ancient Greek philosopher Democritus, who coined the word, ἄτομος or atom, that is, indivisible particle. During the seventeenth and eighteenth centuries, the development of physics concentrated on macroscopic physical properties such as optics, heat, and mechanics, which did not emphasize the atomistic concept. With the development of chemistry and spectroscopy during the nineteenth century, however,

[43]

FIGURE 4.9   Aerial view of the Stanford Linear Accelerator Center. The 22-GeV electron accelerator at this center is the only controlled source of electrons in the world for research in the energy range above 12 GeV. In addition to electrons, this accelerator can provide many beams of secondary particles—pions, muons, kaons, and neutrinos. [Source: Stanford Linear Accelerator Center.]

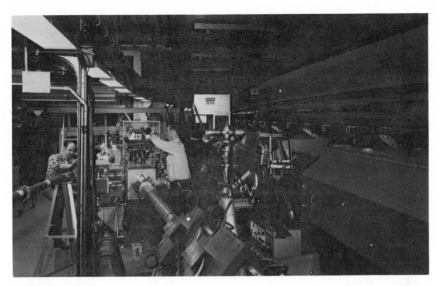

FIGURE 4.10 The AGS magnet ring and linac junction. This 33-GeV accelerator at the Brookhaven National Laboratory is the principal source in the United States for research using protons in the energy range 12 to 33 GeV. The AGS has been undergoing a major conversion that is now nearing completion. The purpose of this conversion is to provide higher-intensity beams of protons and secondary particles and increased experimental capability, thereby making it possible to carry out research of a qualitatively different character than before. [Source: Brookhaven National Laboratory.]

the study of atoms, molecules, electrons, and ions became a most influential part of physics, culminating in the formulation of modern quantum theory in the 1920's. That the atom turned out to be a structured system, made up of electrons and nucleons, was the first in a series of experiences demonstrating the elusiveness of the ultimate elementary constituents of matter. A complete answer to this question is still lacking.

To determine its constituents, it is necessary to break matter apart until it is reduced to an unbreakable minimum. Thus, matter is subjected to extreme conditions by making the components collide with great vigor and break one another apart. This vigor is usually measured in terms of the energy of collision. The energy provided by a flame is great enough to break molecules apart into atoms (fractions of electron volts), but greater energy is needed to strip electrons from the atom (electron volts). After the electrons are stripped off, the bare nucleus of the atom might appear to be elementary unless even greater energy is brought to bear to shatter it.

The magnitude of energy needed to tear the nucleus apart is measured in millions of electron volts, that is, it is of the order of one million times

[45]

greater than that required to tear electrons out of atoms. Sources of particles with such energy are naturally available in the emissions from radioactive nuclei, in cosmic radiations, and in the interiors of ordinary stars. Some of the earliest controlled experiments used radioactive sources, but an attack on the nuclear structure problem over a broad front was not possible until Cockcroft and Walton, Van de Graaff, and Lawrence developed machines (electrostatic generators and cyclotrons) capable of accelerating particles to energies of millions of electron volts.

At this stage, in the 1930's, physics appeared to have attained the goal of determining the ultimate constituents of matter and energy. The elementary particles of which atoms and ordinary matter are constituted were clearly the electron, proton, and neutron. The proton is the positively charged nucleus of the simplest atom—hydrogen. The neutron is its neutral counterpart, discovered by Chadwick in 1932. All other normal nuclei consist of neutrons and protons bound together.

The simple notion that all matter is made up of these three particles was augmented by the concept of the photon as the elementary particle or quantum of electromagnetic energy, a concept that came from the quantum theory of radiation. In this case, the particle is not a constituent of matter but a manifestation of energy released in the transition of matter from a state of higher energy (or mass) to a state of lower energy (or mass). This example shows the interchangeability of matter and energy, a well-known aspect of Einstein's special theory of relativity.

The theory of relativity combined with quantum theory also led to the suggestion that for every particle there should be a corresponding antiparticle, having the same mass but opposite electromagnetic properties, as, for example, the sign of its charge. This hypothesis was first suggested as a result of Dirac's relativistic version of the quantum mechanics of the electron. The positron, which is the antiparticle to the electron, and the first known antimatter, was discovered experimentally by Anderson in 1932.

Another elementary particle that appears only as a result of transitions between different states of matter is the neutrino. The existence of the neutrino (and of the antineutrino) was hypothesized by Pauli in that same era (about 1932) to account for what would otherwise have been gross violations of a fundamental law of nature—conservation of energy and of angular momentum—in the beta decay of radioactive nuclei. It is a particle having no charge or mass, and it has only recently—in very sophisticated experiments—been detected directly. The attempt to detect solar neutrinos is critically important for all of astrophysics. For example, their detection will provide the most direct test of the hypothesis that the sun is generating thermonuclear power (see Figure 4.11).

Although a rather complete picture of the basic constituents appeared to exist in the 1930's, little was known about the properties and structure of

[46]

FIGURE 4.11 The Brookhaven solar neutrino experiment located 4850 ft underground in the Homestake Gold Mine at Lead, South Dakota. Detection is based on observing the neutrino capture reaction $^{37}Cl + \nu \rightarrow {}^{37}Ar + e^-$ in 100,000 gallons of perchloroethylene, C_2Cl_4. The radioactive ^{37}Ar (35-day half-life) is removed from the liquid by a helium gas purge and placed in a small low-level proportional counter to observe the electron capture decay of ^{37}Ar.

[47]

matter at the nuclear level. Even the strong forces responsible for holding nuclei together (against the tendency of the positively charged protons to repel each other and push it apart) were a complete mystery. It was known that these forces were strong and acted over very short distances (short-range forces) of the order of nuclear dimensions—typically 100,000 times smaller than those of the atom. The shape of the potential, if indeed these forces could be described by a potential, and its other features were the subjects of intense experimental and theoretical investigation.

Investigation of the shape of such a short-range potential implied that measurements over very short distances were needed. Such measurements could be made only with particles of high energy, because the characteristic wavelength associated with a particle, in accordance with quantum mechanics, is inversely proportional to its momentum, and very short wavelengths are needed to see structure over very small distances. The energy corresponding to a proton wavelength comparable to the range of nuclear forces is about 10 million electron volts (MeV). Therefore, to study effects over much smaller distances requires proton energies of the order of at least 100 MeV (the energy goes inversely as the square of the distance scale).

For this reason, accelerators in the 100- to 600-MeV range were built in the late 1940's and early 1950's when methods to build such machines (synchrocyclotrons, linear accelerators) had been devised. However, there were many other reasons, too, not the least of them being simple curiosity regarding the open-ended question: How does matter behave under more and more extreme impact? From the viewpoint of particle physics the most important reason was given by Yukawa in 1935 when he proposed a field theory of the nuclear force analogous to electromagnetic theory.

Prior to the study of nuclear phenomena, the forces of nature were assumed to be two in number, gravitational and electromagnetic. Both were known to obey an inverse-square law, with the force, albeit rapidly weakening, extending to arbitrary distances from the force center. It was clear from the very earliest measurements, that the nuclear forces had a quite different character; within an exceedingly short range they were very much stronger than the electromagnetic forces, but they vanished abruptly beyond a distance characteristic of nuclear dimensions. Yukawa showed theoretically that the range of a natural force field was tied inversely to the mass of the characteristic particle or quantum of the field; this finding was in accord with the known zero rest mass of the photon and the effective infinite range of the electromagnetic field. It also suggested that the graviton, the quantum of the gravitational field, should have zero rest mass. To match the short range of the strong nuclear force, however, Yukawa was forced to postulate the existence of a nuclear field quantum of finite mass— roughly one tenth of that of the proton.

[48]

Because nuclear forces are capable of exchanging electric charge between neutrons and protons, the nuclear quanta had to appear in charged as well as neutral forms. Because the mass of this quantum corresponds, according to the relationship $E=mc^2$, to an energy between 100 and 200 MeV, energies in this range are required to produce them. They were first identified in cosmic rays in the late 1940's. They are produced in great abundance by machines giving protons with energies of 400 MeV and higher.* The strong nuclear field quanta are called pi-mesons or pions. Beams of these pions are now an essential tool for elementary-particle and nuclear physics research and are rapidly finding applications in medicine.

A major puzzle emerged after the initial postulation of the pion by Yukawa. By its very nature it was expected to interact strongly with neutrons and protons, hence with nuclei. But when particles of approximately the right mass and charge were first identified in the cosmic radiation, it appeared that, having passed through much of the earth's atmosphere before detection, they interacted very weakly with nuclei. It was finally realized that the difficulty arose because the free pion was itself unstable against decay. In a time of about one hundredth of a microsecond after being produced, the pion transforms spontaneously into a somewhat lighter particle called the mu-meson, or muon, and a neutrino. Because the muon interacts only weakly with nuclear matter, it can survive passage through the earth's atmosphere after being produced from the decay of pions, which result from the interaction of the primary atomic radiation with the matter in the upper fringes of the atmosphere. The muon was initially, and incorrectly, thought to be Yukawa's meson.

This muon is unstable and transforms spontaneously into an electron, a neutrino, and an antineutrino about 1 μsec after production (in essence, it beta-decays). Its observation at sea level simply shows that it is normally produced with very high kinetic energy following the initial cosmic-ray interaction, and the relativistic time dilation (rapidly moving clocks appear to run slow) is such that it survives passage through the atmosphere before decay. Some pions also survive passage through the atmosphere for similar reasons, despite their much shorter lifetime and the much higher interaction probability. Every observation made to date indicates that the muon behaves in every way like a heavy electron. It is not subject to strong interactions (of nuclear strength), it has identical electromagnetic interactions, and, as far as has been determined, it is subject to the same weak interactions that are responsible for beta-decay, except that the muon has

---

* The extra energy is needed in any collision process because the struck particles are not infinitely massive and recoil; in doing so they retain some of the available energy so that less than the total incident energy is available in the interaction itself. In the jargon of collision physics this is referred to as energy *in* the center-of-mass system as opposed to energy of the center of mass.

[49]

its own distinct neutrino (and antineutrino) associated with weak decay processes. It represents a major puzzle; physicists simply have no idea why there should be such a heavy electron or, indeed, whether still heavier electrons remain to be discovered.

Electrons, muons, the two kinds of neutrino, and all their associated antiparticles form a special class of particles called leptons. So far there is no evidence that they interact in a strong way with any form of matter or radiation. Possibly, for this reason, they appear to be truly elementary, that is, irreducible particles.

Particles like the proton, neutron, and pion that do interact strongly, that is, with interaction energies comparable with the potential energies between nucleons, are referred to as hadrons. There are also a great many other kinds of hadron with quite remarkable properties. The first hint of their existence came, again, from cosmic-ray observations, in which a few examples of what are now called hyperons and K-mesons were recognized as unusual events in cloud chambers and emulsions.

The characterization of these particles, their classification, and the relationships between them began to unfold only when the first accelerators in the billon-electron-volt (GeV)* range came into operation. The first such proton accelerator, the 3-GeV Cosmotron at Brookhaven, was planned before the existence of these strange particles was known, but their existence was clearly indicated by the time the Cosmotron was turned on. It was only shortly afterward, as a result of experiments at the Cosmotron, that the principle of associated production of strange particles, which is one of the fundamental concepts in the understanding of strangeness, was well established. Just as in moving from the atomic to the nucleon quantum system it was necessary to introduce a new quantum number, the isotopic spin, to characterize the different charge states of the nucleon—the neutron and proton—so also in going to these elementary-particle systems the addition of yet another label or quantum number, strangeness, was required.

The Cosmotron was built to study nuclear forces at a deeper level than had been possible before and to observe the behavior of matter under more energetic impact. Experiments showed that matter behaved in a totally unexpected manner, thus an entirely new field of investigation emerged. The Bevatron, at the University of California at Berkeley, planned for similar purposes, was completed shortly thereafter. In view of the theoretical reasons for believing that for every particle there is a corresponding antiparticle, the selected energy was sufficiently high (6 GeV) to produce antiprotons. (The first triumph of this theory had been the discovery of the positron.) The mass of the proton is large (about 1 GeV/c^2 in energy units), and antiprotons must be produced in proton–antiproton pairs (just

---

* For giga electron volt; this abbreviation has now replaced the formerly used BeV.

[50]

as positrons are produced in electron–positron pairs), requiring the conversion of 2 GeV of energy. This amount of useful energy (in the center-of-mass system) is produced in the collision of a 6-GeV proton with a proton at rest. Thus, the Bevatron produced antiprotons as well as other known particles.

At this point, it should be clear that the original objective of determining the constituents of matter is not distinguishable from the determination of the possible forms of energy. Not only are the photon and neutrino produced in transitions between energy states of matter, but electron–positron pairs, pions, proton–antiproton pairs, associated pairs of strange particles, and any other form of matter can be created in this way. The distinction between matter and energy has vanished.

Recent Developments in Elementary-Particle Physics

A continuous unfolding of surprises concerning the forms that elementary matter and energy can take characterizes the recent history of this subfield. The intrinsic properties, interactions, reactions, and all aspects of the behavior of these constituents, which are found under more and more extreme conditions, have led to entirely new concepts of the nature of the physical universe. Higher-energy machines, in the 10- to 70-GeV range, that became operative throughout the world during the past 15 years, and the great number of experiments conducted at them, produced many of these revelations. The nature of many of the most important discoveries could not have been anticipated when the machines at which they occurred were being planned.

Among these recent discoveries are the following:

1. Discovery of parity violation in weak interactions (that is, nature distinguishes between right- and left-handedness in these interactions);
2. Discovery of two kinds of neutrino;
3. Confirmation of the idea that the vector part of the weak nuclear interaction is generated in a manner remarkably similar to the generation of electromagnetic interactions;
4. Discovery of a difference between the world of particles and the world of antiparticles, even when the latter is viewed in a mirror (that is, violation of CP invariance), and the associated discovery that there is some aspect of the weak interactions that depends on the direction of flow of time;
5. Elucidation of the structure of neutron and proton in terms of their internal charge and current densities;
6. Discovery that protons appear to have an internal point particulate structure;
7. Realization that electromagnetic properties of particles not only relate

[51]

to the usual massless photons but also involve very massive vector mesons that are subject to strong (nuclear) interaction (reflecting the large masses, these interactions are of very short range);

8. Exploration of the limits of relativistic quantum electrodynamics to a distance of the order of 10^{-14} cm;

9. Opening of the field of hadron spectroscopy (Table 4.1) and discovery of the underlying $SU(3)$ symmetry relating the hadrons;

10. Realization of phenomenological theories for dealing with relativistic reactions between hadrons and interpreting them, especially at very high energy.

The report of the Panel on Elementary-Particle Physics in Volume II discusses a number of these findings in greater detail. What emerges is a

TABLE 4.1   Stable Hadrons

Name	Symbol	Electric Charge	Mag. of mc^2 (MeV)	Spin	Parity	Isotopic Spin	Strange-ness
Pion	$\pi^\pm$	$\pm e$	140	0	neg.	1	0
	π^0	0	135				
K-meson	K^+	$+e$	494	0	neg.	½	$+1$
	K^0	0	498				
$\overline{K}$-meson	K^-	$-e$	a	a	a	a	-1
	$\overline{K}^0$	0					
Nucleon	p	$+e$	938.3	½	pos.	½	0
	n	0	939.6				
Antinucleon	$\overline{p}$	$-e$	a	a	a	a	0
	$\overline{n}$	0					
Lambda	Λ	0	1116	½	pos.	0	-1
Antilambda	$\overline{\Lambda}$	0	a	a	a	a	$+1$
Sigma	$\Sigma^\pm$	$\pm e$	1197	½	pos.	1	-1
	Σ^0	0	1192				
Antisigma	$\overline{\Sigma}^\pm$	$\pm e$	a	a	a	a	$+1$
	$\overline{\Sigma}^0$	0					
Cascade	Ξ^-	$-e$	1321	½	pos.	½	-2
	Ξ^0	0	1314				
Anticascade	$\overline{\Xi}^+$	$+e$	a	a	a	a	$+2$
	$\overline{\Xi}^0$	0					
Omega minus	Ω^-	$-e$	1672	½	pos.	0	-3
Antiomega minus	$\overline{\Omega}^+$	$+e$	a	a	a	a	$+3$

a Masses and indicated quantum numbers are the same for particle and antiparticle of opposite charge.

[52]

continuing progression of new discoveries and new surprises that represent salients into entirely virgin territory in man's understanding of the nature of space, time, and the basic laws that govern all natural phenomena. This push into the unknown is one of the greatest adventures of the human intellect; but it is much more. It is a continuing challenge to further exploration and fuller understanding of natural phenomena. It commands the devotion and attention of some of the most intellectually gifted scientists of this age.

Interactions with Technology

Successes in elementary-particle physics have depended heavily on the discoveries and developments made in almost all subfields of physics and in engineering. Many new technical developments were needed. The demands of this subfield presented a challenge to technology as great as any offered by either space research or military requirements. As the magnitude of the energy needed for exploration has increased, the size of the required apparatus has increased enormously and so has the degree of ingenuity required both to reduce costs and to provide methods for making the necessary measurements.

The development of new concepts in machine technology made the building of giant machines feasible. The particle beams of very high energy produced by these machines had to be manipulated in a way that required the development of intricate beam-transport devices, including huge bending magnets, magnetic lenses, electrostatic and radio-frequency velocity selectors, computer monitoring and control of beams, and the like. The detection and measurement of particles at high energy and intensity require special detection devices such as bubble chambers, Cerenkov and other counters, spark chambers, and wire chambers, all in combination with on-line computers (see Figure 4.12). Some of these devices must be capable of measuring time intervals of a few billionths of a second or of handling millions of particle events in seconds. Measurements on the photographic film on which information is stored has led to the development of automatic pattern-recognition devices that have many other applications. Computer and special methods had to be developed to perform the many intricate calculations that are needed before the physics can be recognized in the enormous mass of data emerging from an experiment. This task has strained the capabilities of the largest computers. Without such methods, the emergence of hadron spectroscopy, for example, would have been impossible.

Big superconducting magnets (see Figure 4.13) have been designed and built to provide the large magnetic fluxes needed to deflect high-energy particle beams at a reasonable cost, because the power requirements for some individual magnets would be as high as 10 MW. More recently,

[53]

FIGURE 4.12　The 12-ft-diameter bubble chamber that operates in conjunction with the 12-GeV zGs proton accelerator at the Argonne National Laboratory. The chamber, which can be filled with hydrogen or deuterium, is the largest particle detector of its type in operation in the world today. The early work with the chamber has concentrated on neutrino physics experiments but will soon make use of more conventional beams, for example, pions, kaons, and antiprotons. [Source: Argonne National Laboratory.]

[54]

FIGURE 4.13. The 184-in. superconducting magnet (18 kG) built for the Argonne National Laboratory 12-ft bubble chamber. The 100-ton magnet consumes only 10 W of power, much less than the several megawatts required for conventional magnets of the same field strength. This is the largest superconducting magnet operating in the world. [Source: Argonne National Laboratory.]

alternating current and radio-frequency superconducting systems have become feasible and should soon make possible the building of accelerators at higher energy or with an improved duty cycle * without an increase in size or power requirements. Such concepts as the electron ring accelerator may also make it possible to go to much higher energy at reasonable cost.

This dependence of high-energy physics on technology and engineering frequently stretches the capabilities of existing technology to the utmost, requiring innovations and extrapolations that go well beyond any present state of the art. Because the resulting technological developments have implications much broader than their use in particle physics, all technology

---

* In lower-energy electrostatic accelerators, for example, the stream of particles in the beam current is constant in time, that is, it is a dc beam. In the larger accelerators, however, as a consequence of the accelerating mechanisms, the beam particles occur in bursts of greater or lesser length, with intervening periods in which there are no particles. The duty cycle is a measure of the extent to which a given accelerator approaches a dc beam character.

benefits from the opportunity to respond to this pressure. New technical developments occur sooner—sometimes much earlier—than they would in the absence of such pressure, and they often present new engineering opportunities, unrelated to high-energy physics, that can be exploited immediately. Examples of such engineering developments are manifold: large volume, very-high-vacuum systems; sources of enormous radio-frequency power; cryogenic systems; large-scale static superconducting magnets; variable-current superconducting systems; pattern-recognition devices; very fast electronic circuits; and on-line computer techniques. In this sense, elementary-particle physics has had a major impact on technology, but the effect has been an indirect one resulting from the urgency of the research requirements rather than the results of the research.

The history of physics offers many examples of the discovery of elementary particles, or properties of elementary particles, that have had extremely important applications. These examples are the basis of a belief that it will happen again. Examples of important innovations in technology based on particles whose existence is yet to be established have been suggested. Of course, physicists are on tenuous ground here, because such speculations could be based on false premises. That is why more research is needed—to determine the validity of the premises. If they are not correct, past experience suggests that the results could be completely unexpected and might have greater impact on technology than can be imagined at the present time.

It may be helpful to include an example of such a speculation suggested by recent discoveries in particle physics. The discovery of a multitude of particle resonances led to the development of hadron spectroscopy. This study has revealed many remarkable regularities in the relationships among the hadrons, which, in turn, have been classified in terms of several parameters that take on a very limited set of values and are called internal quantum numbers.

The situation is analogous to that which occurred in the nineteenth century in connection with the concept of chemical valence, which also can be regarded as an internal quantum number of the atom; the regularities exhibited a pattern that led Mendeleev to construct the periodic table of elements, which not only systematized existing information but also had predictive power. Every gap in the table was later filled by a chemical element whose existence was unknown at the time that the table was constructed.

In regard to the regularities of hadron spectroscopy, the internal quantum numbers have been associated with the symmetry patterns (representations) related to the group $SU(3)$ (long familiar in group theory) of simple three-dimensional unitary transformations. These patterns also can be used to predict the particles that should exist and some of their properties. The predictions thus far have been remarkably successful, although there are some contradictions and controversies that must be resolved.

The remarkable successes of the $SU(3)$ scheme, for which Gell-Mann was awarded the 1969 Nobel Prize in physics, have led to the type of situation of which physicists dream. In the case of the particle, then unknown, the theoretical prediction was remarkably specific. It included the mass, spin, charge, and all observable characteristics of the particle. Moreover, substantiation of the entire theory depended on its existence. Thus, discovery of the omega minus, with all the predicted properties, was a major vindication of this approach and the underlying understanding of fundamental phenomena on which it was based. Even more recently, the antiparticle, the antiomega minus, has been observed after strenuous effort by a group at Berkeley (Figure 4.14). Its discovery fills all the existing gaps in the $SU(3)$ ordering; however, this does not preclude the possible discovery of deeper symmetries or a broader ordering involving yet unknown species of particles.

The physics of the periodic table of the chemical elements was ultimately elucidated in terms of the electron structure of the atom, the quantum mechanics of the electron, the spin of the electron, the Pauli exclusion principle, and the like. To make an analogy with the situation in particle physics today, one could ask how the periodic table of elements might have been interpreted if the quantum mechanics of elementary particles had been known but the existence of the electron had not. It is not too difficult to imagine that someone could have come to the conclusion, on the basis of the periodic table of the elements, that the atom consisted of particles of spin one-half, satisfying Fermi-Dirac statistics (that is, the Pauli exclusion principle) and moving in a central field of force. This conclusion would not have been sufficient to characterize the electrons in atoms completely; but, in the context of the analogy, it would have been cause for subjecting atoms to severe collisions involving energies of kilovolts, which would have been high energy in the nineteenth century, to knock some of the particles (electrons) out of the atom for further study.

The corresponding problem in particle physics is the explanation of the observed $SU(3)$ symmetry of the hadrons. It has been suggested that $SU(3)$ may represent the structure of the hadrons as composites of ultimate particles of three distinct kinds, which have been given the name quarks. All mesons would be composed of appropriate pairs consisting of one quark and one antiquark each; the baryons, which are particles like the proton and neutron, would consist of three quarks.

To keep the model simple, that is, to limit it to just three kinds of quark and their associated antiquarks, it is necessary that the electric

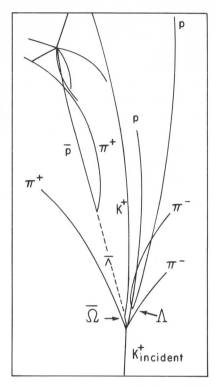

FIGURE 4.14.   Almost all of the antibaryons have been observed in experiments, the most recent being the antiomega particle ($\overline{\Omega}$), in 1970, by a group from the University of California, Berkeley. The bubble chamber event is shown in the above figure. The experiment involved was a study of the K⁺d interaction at 12 BeV/c carried out in the 82-in. Stanford Linear Accelerator Center bubble chamber. The production reaction is $K^+d \rightarrow \overline{\Omega}\Lambda\Lambda p\pi^+\pi^-$, and the decay is $\Omega \rightarrow \Lambda K^+$. [Courtesy of Gerson Goldhaber.]

charges of these particles occur in units of one third of the elementary charge, *e*. There is, of course, no evidence whatsoever that this is actually possible. The three different kinds of quark would necessarily have charges of $\pm e/3$ and $2e/3$. The antiquarks would have charges of the opposite sign. The quark idea leads to simple interpretations of many of the known characteristics of hadrons, but the particles have other peculiarities and suggest many unanswered questions. That very massive quarks can combine, at least theoretically, to give much lighter resulting particles is simply a reflection of the correspondingly large binding energies that would be involved.

The key question is: Does the quark really exist? The energy of collisions between hadrons required to produce quarks depends on the as yet unknown quark mass, and it may be very high. They have not been detected at existing accelerators, either because the available energies were inadequate or because the quarks are a mathematical fiction, serving only as a simple and graphic representation of something even deeper in the theory. Despite this situation, one of the earliest experiments with the new National Accelerator will be a search for the quark. There is an elegance and an economy about this quark hypothesis that has great aesthetic appeal. (See Plate 4.II.)

The connection with technology is highly speculative and depends on the anticipated stability of the hypothesized quark. This stability implies that there may be ways to store quarks after producing them at an accelerator. The storage of quarks would correspond to the storage of enormous amounts of energy, and this energy could be released in a controlled fashion by allowing quarks and antiquarks to recombine at the desired rate.

Because electric charge is conserved in all reactions, there is no way for a quark to disappear in ordinary matter; therefore, it is stable not only in vacuum but also in the presence of matter. The making of exotic atoms by binding a proton to a negatively charged quark is conceivable. Since the quark is expected to be the more massive, the proton would play a role similar to that of the electron in an ordinary atom. Possibly, exotic atoms could form exotic molecules. One can immediately speculate that such exotic atoms and molecules might be used to catalyze thermonuclear reactions between protons and deuterons, because they would provide a mechanism to juxtapose the nuclei in spite of the repulsive forces caused by their positive electric charges. Relatively few quarks should be needed, since their release for reuse at each thermonuclear interaction would be expected. Thus, it is possible to imagine a controlled source of thermonuclear power consisting of a vessel of deuterons into which just enough quark impurity has been introduced to produce power at the desired rate. Although this speculation may seem farfetched, thermonuclear catalysis by muons in an analogous process already has been observed; only because muons are unstable are they useless as a practical source of energy.

It is quite possible that quarks do not exist. But the investigation of this question almost certainly will lead to the discovery of an even more fascinating world.

Impact on Other Sciences

Since elementary-particle physics has its origin in nuclear physics, there is a particularly close connection between these two subfields. They depend on one another in many ways. The research results of particle physics having the greatest direct bearing on nuclear physics are those that throw light on the nature of nuclear forces and those that provide methods for the study of nuclear structure. (See the following section on Nuclear Physics.)

Not only are there results of research that are of interest to both sub-fields, but there is a substantial overlap in the instrumentation, so that each learns from the other. There also has been a strong mutual impact in regard to methods for handling data, use of on-line computers, and the like.

High-energy physics also plays a special role in relation to astrophysics, because the astrophysicist is concerned with matter under the most extreme conditions. At sufficiently high temperatures the exotic and unstable particles produced by accelerators will exist in thermal equilibrium with matter and influence the equation of state of matter under these extreme conditions as well as determine the reactions that occur. Thus a knowledge of all the elementary particles and how they behave is essential to a full understanding of interstellar matter under some conditions. There is also an important relationship with the work in space-radiation physics. The cosmic rays were the first source of high-energy particles, and the study of these particles in space makes full use of both existing knowledge of their properties and instruments and techniques developed for their study.

In subfields other than nuclear physics and astrophysics, the connection is less direct. New instruments developed for particle physics are useful in many disciplines. For example, various pattern-recognition devices for scanning and measuring film from bubble chambers and spark chambers are valuable for a variety of applications in biological research, two of which are chromosome counting and measuring cross sections of nerve bundles. Image intensifiers, developed to photograph particle tracks in scintillators, have proved useful for quantitative observation of chemical luminescence in biological systems. Another important medical application has arisen in connection with the development of techniques for making and handling thin but very tough plastic films as support for multiwire particle detectors, plastic plumbing for very rugged thin-walled targets, and other plastic systems (Figure 4.15). Elementary particles could open new fields of research in medicine. For example, there is reason to believe that irradiation of tumors with pion beams may have therapeutic value because of the particular properties of the pion. To establish the validity of this idea requires extensive research on the biological effects of pion beams.

Theoretical methods of one branch of physics are usually applicable in others. The history of physics has been characterized by the unity of the theories (as discussed in the final section of this chapter). Elementary-particle theory is no exception. The methods of quantum field theory, developed to answer fundamental questions concerning particles, have

Combining Quarks

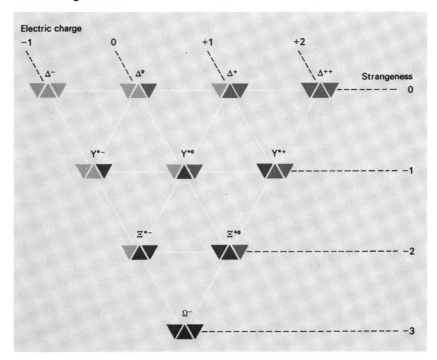

PLATE 4.II(a) Combining quarks. Three kinds of object, shown as different patterned triangles, can be arranged into ten possible groups of three. In the Eightfold Way, each combination of three fictitious objects, called quarks, makes a different baryon. [Source: *Science Year. The World Book Science Annual.* Copyright © 1968, Field Enterprises Educational Corporation.]

[61]

One Baryon Family

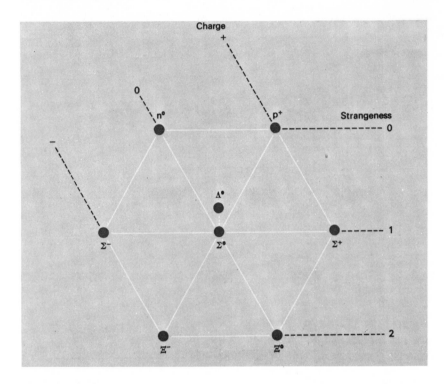

FIGURE 4.II(b) One baryon family. Complex combinations of the three quark species form other families of baryons, according to The Eightfold Way. The neutron and proton are in the top row. [Source: *Science Year. The World Book Science Annual.* Copyright © 1968, Field Enterprises Educational Corporation.]

FIGURE 4.15 A preliminary version of a small, inexpensive ($15) artificial kidney, which has been developed at the High Energy Facilities Division of Argonne National Laboratory. It has been used successfully by nine patients at an Illinois Veterans Hospital. It has small enough volume ($6 \times 2 \times 2$ in.^{3}) to be suitable for a child, and it is hoped that it can be made inexpensive enough to be used daily. The basic configurations of these dialyzer designs are also being explored for their potential as membrane oxygenators for heart–lung machines. [Source: Argonne National Laboratory.]

proved to be applicable to problems in many other subfields such as the nuclear many-body problem, many-body problems in condensed matter, superconductivity theory, and statistical mechanics. Methods for treating resonance reactions, a field that had its beginning in atomic physics and flowered in nuclear physics, have required deeper investigation to understand their meaning in the relativistic processes that occur at high energy. The result has been more general methods and insights into scattering processes in general, with applications whenever they occur.

The methods for dealing with the general problem of relativistic reactions have been the subject of intense investigation, with results that have widespread use and add to current understanding of all similar physical phenomena. However, not only the methods but also the content of these theories can be of great importance in other subfields of physics. Just as the full panoply of elementary particles and their excited states can exist in stellar interiors, so in atoms and molecules such particles exist in virtual form for minuscule instants of time and thus have a tiny influence on atomic and molecular properties that will eventually be detected when measurements become sufficiently precise. Therefore, it is possible that the limits of the known theory of electrons and electromagnetism will be passed by high-precision measurements of an atomic nature before they have been reached by high-energy experiments. The two approaches are complementary, and both have been pursued vigorously, but the only certainty is that a limit will eventually be reached, a limit such that even the theory of the electronic structure of atoms will require modification.

Although there are many specific examples of the impact of elementary-particle physics on other sciences, especially on other subfields of physics, it is not these details that represent its most significant contribution. This contribution is associated with the whole flavor and character of scientific exploration and discovery. The entire scientific organism flourishes when its major fields and subfields are flourishing and feeding the whole. In different fields, and in different subfields of physics, the nature and style of the work varies greatly, but all science gains strength when any subfield contributes in a substantial way to the overall store of knowledge or under-standing. Certainly all of physics gains from any significant new discovery, idea, information, or theory in any of its major subfields.

Of course, there is no way to guarantee the unexpected, and the pursuit of it is a gamble. But the gamble has been paying off handsomely, and there is no reason to believe that it will not continue to do so.

Distribution of Activity

Elementary-particle physics grew from academic research in atomic, nu-clear, and cosmic-ray physics, and, in spite of its enormous development in scale, its roots have remained in the universities. The early accelerators and the devices for studying the behavior of particle beams emerging from them were built at the universities. A few accelerators at universities con-tinue to be used for elementary-particle physics experiments, but the emphasis has shifted to much greater energies and more complicated equip-ment than can be managed within the usual university framework. There-fore, the accelerators at these higher energies have been built at National Laboratories, some of them managed by single universities and others by consortia of universities (see Chapter 9). Although each of these labora-tories has its own in-house research activity, by far the larger part of the research in the National Laboratories is conducted by groups of professors, research associates, and students from universities throughout the country.

[64]

Table 4.2 lists the operating high-energy accelerators of the world. The Cosmotron, a 3-GeV proton accelerator at Brookhaven, was shut down several years ago, and, as noted in the table, the Princeton Particle Accelerator (PPA) has been shut down as an elementary-particle physics laboratory. These shutdowns resulted from funding limitations, not from any significant obsolescence of the scientific value of the equipment.

The 200–500-GeV proton accelerator at the National Accelerator Laboratory is nearing completion. It is possible that the first experiments will be carried out there in mid-1972. Already initial design work has begun

TABLE 4.2 Accelerators Operating above 1.5 GeV [a]

Accelerated Particle	United States		Western Europe		Soviet Union	
Proton	PPA [b]	3 GeV	Saturne	3 GeV		
Proton	Bevatron	6.2 GeV	Nimrod	7 GeV	ITEP	7.5 GeV
Proton	ZGS	12.7 GeV				10 GeV
Proton	AGS	33 GeV	CERN-PS	28 GeV	Serpukhov	76 GeV
Electron			Bonn	2.3 GeV	Kharkov	2 GeV
Electron	CEA [c]	6.3 GeV	NINA	5 GeV		
Electron	Cornell	10 GeV	DESY	6.2 GeV	Yerevan	6 GeV
Electron	SLAC	21 GeV				

[a] Abbreviations, names, and locations shown in table:

United States

PPA	Princeton Particle Accelerator, Princeton, N.J.
Bevatron	Lawrence Berkeley Laboratory, Berkeley, Calif.
ZGS	Zero Gradient Synchrotron, Argonne National Laboratory, Argonne, Ill.
AGS	Alternating Gradient Synchrotron, Brookhaven National Laboratory, Upton, Long Island, N.Y.
Cornell	Cornell University, Ithaca, N.Y.
CEA	Cambridge Electron Accelerator, Harvard University, Cambridge, Mass.
SLAC	Stanford Linear Accelerator Center, Stanford, Calif.

Western Europe

Saturne	Commissariat à l'Énergie Atomique, Saclay, France
Nimrod	Rutherford Laboratory, Chilton, Berkshire, England
CERN-PS	Proton Synchrotron, CERN, Geneva, Switzerland
Bonn	Physikalisches Institut, Bonn, Germany
DESY	Deutsches Elektronen-Synchrotron, Hamburg, Germany
NINA	Daresbury Nuclear Physics Laboratory, Daresbury, England

Soviet Union

ITEP	Institute of Theoretical and Experimental Physics, Moscow
Serpukhov	Institute of High Energy Physics, Serphkhov
Kharkov	Physical Technical Institute, Kharkov
Yerevan	Institute of Physics (GKAE), Yerevan, Armenian SSR

[b] Shut down as an elementary-particle physics facility at end of fiscal year 1971.

[c] Being used only as an intersecting-beam device.

that will extend the capability of this facility to 1000 GeV through the addition of a ring of superconducting magnets immediately above the present magnet ring (Figure 4.16).

At the present time, there are approximately 50 U.S. universities heavily involved in elementary-particle physics. In addition, the number participating to some degree, with hope of greater involvement, is about 125.

The sociology of high-energy physics has become rather complicated because the determining factors in the research and training activities in the universities are the decisions made concerning the experiments to be carried out or the ancillary equipment to be developed at the large accelerators at the National Laboratories. In order that the judgments and needs of the participating research community may be taken into account, an elaborate system of advisory committees and corporations formed by con-

FIGURE 4.16   Aerial view of main accelerator at the National Accelerator Laboratory. This 200–500-GeV proton accelerator will be for some years the only controlled source of protons in the world for research in the energy range above 80 GeV and the only one in the United States above 33 GeV. In addition to proton beams, the accelerator will provide many beams of secondary particles—neutrinos, pions, kaons, photons, and antiprotons. [Source: National Accelerator Laboratory.]

sortia of universities has come into being. The arrangements are tailored to each laboratory and its special problems and needs. They seem to work rather well, at least as judged by the success of the research programs (see also Chapter 9).

There are approximately 1700 PhD physicists and engineers who can be identified as working on projects supported by federal particle-physics funds. Of these, about one third are theoretical physicists, the rest, experimental. This figure probably does not include a substantial number of theoretical particle physicists who do not receive federal support.

The number of graduate students at the thesis level working on programs supported by federal particle-physics funds is about 1100. A substantial number of students supported by other funds should probably be added to this number. Well over 300 PhD's annually are being granted in this subfield, and more than half of them are based on theses in theoretical physics.

About half of these particle physics PhD's have gone into other subfields after receiving their degrees. Those who remain usually spend from two to four years as research associates, or in equivalent temporary postdoctoral

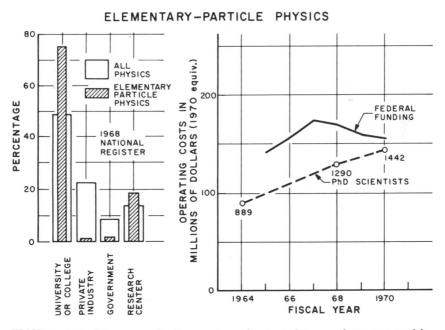

FIGURE 4.17 Manpower, funding, and employment data on elementary-particle physics, 1964–1970.

[67]

positions, in either a university or a National Laboratory. Because of the current limitations on permanent job opportunities, many research associates are extending their appointments beyond even the fourth year, and the number of PhD's leaving the subfield is increasing (see Appendix 4.B).

Elementary-particle physicists comprise about 10 percent of the physics PhD population, as reported in the 1970 National Register of Scientific and Technical Personnel. They work principally in academic institutions. One fifth were employed in research centers (for the most part, National Laboratories). The number of PhD's in the subfield has grown at an annual rate of 10 percent since 1964 (see Figure 4.17). Approximately 33 percent of the federal funds for basic research in the various subfields of physics in 1970 were allocated to the support of research in elementary-particle physics; there is essentially no direct industrial support (see Figure 4.18).

Problems in the Subfield

Funding The funding of high-energy physics has developed serious inconsistencies in the past few years. Although the present capability, measured in terms of both equipment and manpower, probably provides the United States with the greatest potentiality for research in this subfield of any country in the world, the bleak funding pattern that has characterized federal budgets of recent years seems likely to lead to rapid dissipation of

ELEMENTARY-PARTICLE PHYSICS

PHYSICS MANPOWER IN 1970

N = 36,336

FUNDS FOR BASIC
RESEARCH IN PHYSICS IN 1970

FEDERAL

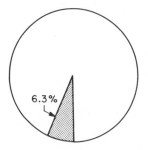
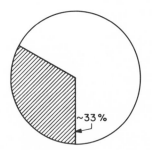

FIGURE 4.18 Manpower and federal funding in elementary-particle physics in 1970.

this strength. Accelerators are being utilized at less than 75 percent of capacity because of financial stringencies and are facing even further cuts. University groups are finding it increasingly difficult to obtain the funds needed to mount experiments that will take advantage of the available facilities.

The total funding of operations and equipment of the principal accelerator laboratories has been decreasing in absolute dollars at a time when one new major accelerator (SLAC) has come into operation, when a major improvement program increasing the capability of the AGS is being completed, and when university groups from all parts of the country are trying to prepare for experiments at the National Accelerator Laboratory (NAL).

As NAL operations get under way, it is essential that incremental funding be provided in the total operating and capital equipment budgets for all accelerators. Otherwise, the exploitation of NAL, which offers the most important opportunities in this subfield, will certainly force the demise of some of the lower-energy accelerators and a drastic reduction in the level of activity of others. Under these conditions much of the important work that remains to be done throughout the spectrum of energies below the 200 GeV available at NAL could not be carried out. This situation would seriously weaken elementary-particle physics research in this country.

Manpower The employment problem for theoretical particle physicists appears to be even more serious than it is for other physicists. The large number of such theorists produced in recent years and their high degree of specialization are often given as the causes of this difficulty. This narrow specialization is already an indication that the student of particle theory has been allowed to choose unwisely, because real success in any part of physics requires more breadth, and both great breadth and depth of perspective are required for a significant contribution, especially in theoretical particle physics.

It is imperative that university groups in elementary-particle theory act to discourage all but the most able potential students and that even these should be made fully aware of the employment problems and restricted career opportunities that now appear probable. University groups have a responsibility to expose their most brilliant and able students to the opportunities in all subfields of physics, particularly under present circumstances in which such students will be needed as effective partners in national attacks on major problems affecting society now and in the future. Because a sizable fraction of the most able students are attracted to elementary-particle physics, a correspondingly heavy responsibility rests on those who are now active in this subfield to provide objective and realistic advice concerning these career opportunities.

[69]

First we must inquire whether the elements are
eternal or subject to generation and destruction;
for when this question has been answered their
number and character will be manifest.

ARISTOTLE (384–322 B.C.)
On The Heavens, Book III, 304

NUCLEAR PHYSICS

Introduction

Nuclear physics includes the study of the structure of atomic nuclei and
their interactions with each other, with their constituent particles, and with
the whole spectrum of elementary particles that is now provided by the
very large accelerators. The nuclear domain occupies a central position
between the atomic range of forces and sizes and those of elementary-
particle physics, characteristically within the nucleons themselves. As
the only system in which all the known natural forces can be studied
simultaneously, it provides a natural laboratory for the testing and extending of many of the fundamental symmetries and laws of nature.

Containing a reasonably large, yet manageable, number of strongly
interacting components, the nucleus also occupies a central position in the
universal many-body problem of physics, falling between the few-body
problems, characteristic of elementary-particle interactions, and the extreme
many-body situations of plasma physics and condensed matter, where
statistical approaches dominate; it provides a rich range of phenomena
and the hope of understanding these at a microscopic level.

Although still a relatively young science, nuclear physics has already
had a profound effect on both peace and war and on man's view of his
universe and himself. The release of nuclear energies in fission reactors
holds high promise of providing the energies for civilization until fusion
systems can be perfected or, perhaps, indefinitely; radioactive techniques
have already had a major impact on technology and on both clinical and
research medicine and biology.

Historical Background

The experiments of Rutherford and his colleagues at the beginning of this
century established the basic structure of the atom and located the nucleus
as a massive, positively charged, and very much smaller (by a factor of
some 100,000) entity at the atomic center. Over the next four decades, the
gross characteristics of nuclei were established—their sizes; their basic
components, the neutron and proton; the characteristic energies by which
they were bound together (some millions of times those in atomic systems);

and the nature of the radiations that were spontaneously emitted by some heavy nuclear species. Much of this information came from the study of collisions between nuclei and the nuclear projectiles as more of these became available at increasingly higher energies. In the beginning only the naturally occurring nuclear emanations—the alpha particles (helium nuclei) so dear to Rutherford—were available; the development of the early accelerators by Cockcroft and Walton, Van de Graaff, Lawrence, and many others led to a new era of experimentation in that, for the first time, the experimenter could choose and vary the nature and energy of his projectile, albeit within rather strict instrumental limits, instead of making do with what nature had provided.

Indeed, the development of accelerators of ever-increasing power and sophistication has been one of the main themes of nuclear technology; along with this has gone the development of nuclear radiation detectors of increasing sensitivity and resolution adequate to the challenge of sorting out the products and results of the accelerator projectile-induced nuclear interactions.

By 1949, a considerable body of knowledge had been accumulated on the simpler properties of the 300 or so stable nuclei found in nature and of the perhaps 500 additional radioactive isotopes that had been produced from these using either accelerators or the newly developed nuclear fission reactor. Apart from its energy-generation capabilities, the reactor for the first time made it possible to produce many of these radioactive species in relatively large quantities for study and application. While only a modest amount of information was available concerning any given nucleus, certain striking regularities already were becoming evident as the number of neutrons or protons was varied. These regularities led Mayer and Jensen to propose their remarkably successful shell model of the nucleus, quite analogous to the Bohr model of the atom in that the single nucleons were assumed to move almost independently in well-defined orbits in a central potential well.

This model proved to be a key idea in the understanding of nuclear structure and dynamics. But it was far from obvious! The success of the Bohr-Wheeler liquid-drop nuclear model in explaining the gross characteristics of nuclear fission in 1939 rested on the basic assumption that the nuclear constituents, the nucleons, interacted so strongly that the mean free path between interactions was much less than the diameter of the nucleus itself. This assumption was in apparent direct conflict with the well-defined orbital motion of the shell model, and it was not until 1956 that Weisskopf explained this apparent paradox as a consequence of the operation of the Pauli principle within the nucleus.

In the 1950's, a phenomenal increase in the amount of nuclear information available occurred as new accelerators, detectors, and instrumentation

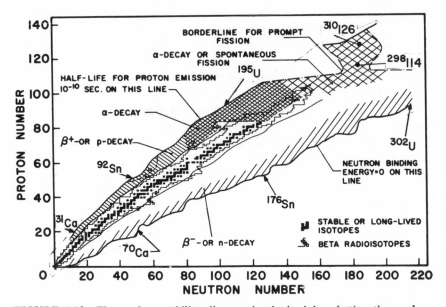

FIGURE 4.19 The nuclear stability diagram is obtained by plotting the nuclear binding energy as a surface deformed by the number of protons (Z) and the number of neutrons (N) present. For low N and Z the most stable species are those having $N=Z$; but with increasing Z, the electrostatic repulsion between the protons forces the stability toward nuclei where N is greater than Z. The black squares represent the nuclei that are stable in nature; this simply means that their lifetime against spontaneous decay is long compared with the lifetime of the solar system. There are some 300 in all. The region outlined with a light line is that which has been explored thus far in nuclear physics; it contains some 1600 different nuclear species (isotopes). The outer solid lines define the region in which it can be calculated that nuclear species should be stable against instantaneous decay via the strong nuclear forces. The lower line is the so-called neutron-drip line in that any species formed below it spontaneously and instantaneously emits neutrons and moves to the left until it reaches the line and the boundary of stability. Emission of tightly bound alpha particles (helium nuclei containing two protons and two neutrons each) competes favorably with direct proton emission at the upper boundary, and spontaneous fission limits the stability region for very heavy nuclei. $B_f=0$ indicates zero binding against such spontaneous fission as does $B_n=0$ against neutron emission. Two of the postulated islands of stability far beyond the natural range are indicated at $Z=114$ and $Z=126$. It is clear from such a figure that a vast range of nuclear species remains to be explored. While the most stable uranium isotopes have masses of 235 and 238, respectively, as shown here, uranium isotopes ranging in mass from 195 to 302 would be expected to be stable against instantaneous breakup.

derived from wartime technology became operational. Not only was much more known about each nucleus, but also many more nuclear species were produced and studied as the new facilities permitted forays from the valley of nuclear stability, which runs diagonally through any map of the nuclear domain wherein the number of neutrons is plotted against the number of protons. Such a map is shown in Figure 4.19. The total number of known nuclear species increased to about 1200, and, instead of

FIGURE 4.20 The first of the Emperor class of electrostatic Van de Graaff accelerators, installed in the Arthur Williams Wright Nuclear Structure Laboratory at Yale University. This accelerator is the largest of the tandem configurations in research use and is the first to make available all the nuclear species to precision study. Terminal potentials in excess of 11 MeV have been obtained on the machine. The large pressure vessel in the background contains the electrostatic accelerator structure; the magnetic systems in the foreground are part of the beam energy control and distribution system.

typically three or four quantum energy levels known for each, numbers such as 20 to 50 became typical.

Yet another wave of innovation and development in the instrumentation of nuclear physics took place in the 1950's. The large higher-energy electrostatic accelerators, for the first time, made all nuclei accessible to precision study with a very broad range of projectiles (see Figure 4.20); the semiconductor nuclear detectors improved the attainable energy resolution by factors between 10 and 100 (see Figure 4.21), and on-line computer techniques not only produced striking improvement in the quality of the available data but also made it possible to digest and analyze the flood of new results efficiently and effectively.

These instrumental innovations were reflected in a massive increase in the scope and quality of information about nuclei. Not only did the number of known nuclear species grow to some 1600, but the information on the quantum structure of each of these increased by orders of magnitude. In addition, entirely new insight into the nature of nuclear interactions and dynamics was gained—that is, how energy and linear and angular momen-

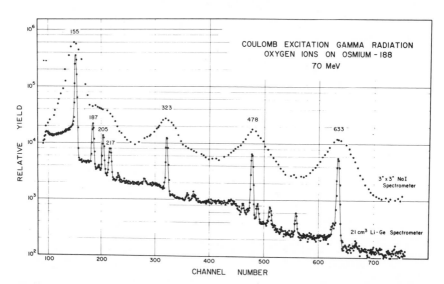

FIGURE 4.21   Comparison of the gamma radiation spectra, from Coulomb excitation of osmium-188 nuclei by a 70-MeV beam of oxygen ions, as measured with a sodium iodide spectrometer and with a lithium–germanium semiconductor spectrometer. In the region of channel number 200, for example, three isolated transitions appear in the latter spectrum that are entirely masked in the former. This is typical of the enormous improvement in resolution that has been attained with these new detectors.

tum are absorbed or emitted by nuclear systems; how nucleons or clusters deform, rotate, and vibrate; and how entirely new nuclear species might be created.

As a result of these studies, the fundamental modes of nuclear excitation are being uncovered and their microscopic structure analyzed. Crucial to this progress was the availability of a large number of complementary means of probing nuclear excitations, of many well-understood ways of building up an excitation, and of examining the way in which the fundamental modes vary systematically from nucleus to nucleus. It is perhaps particularly characteristic of nuclear physics that some of the most far-reaching discoveries came not from increasingly detailed study of a given nuclear system but rather from the recognition of certain underlying systematics traceable from nucleus to nucleus. A very broad effort—broad in kinds of facility and program—was required and has grown in response to this need.

Major progress has been made in understanding the general characteristics of the lower-lying nuclear excitations and nuclear dynamics. The shell model, with appropriate modifications, continues to provide the framework for a physically understandable picture of nuclear structure and motions (see Figure 4.22). To do so it had to encompass phenomena that at first seemed sharply at variance with the concept of single particles orbiting their separate ways. The collective excitations that were discovered and explored in the decades after its initial development clearly pointed to the cooperative motions of many nucleons. Sets of excitations that could only be interpreted as vibrations or rotations of large fractions of the nuclear charge or mass were especially striking testimony to this. The analysis of these collective modes in terms of the shell model appears qualitatively possible and, in some cases, also quantitatively possible, though complex. The new, unified shell model puts the single-particle and collective aspects together by including the residual interactions between the valence nucleons that tie their separate motions together and by recognizing that the assumed spherical central potential, or core, which defined the orbits in the initial shell model, is an oversimplification that, in large regions of the periodic table, must be replaced by spheroidal or ellipsoidal shapes. Besides forming a most useful physical model of the nucleus, this enlarged shell model is also a quantitative bridge between experimental findings and a deeper theory.

Development of a deeper theory has begun. The fundamental question of nuclear theory always has been how to connect the forces between the individual neutrons and protons of the nucleus and the observed nuclear processes. The work of the last 15 years already has achieved more than qualitative success in connecting these forces with both the

[75]

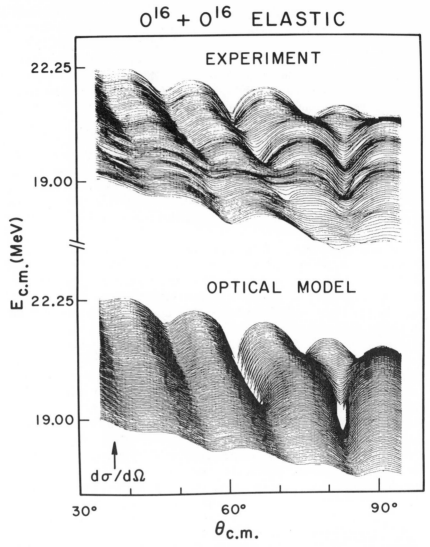

FIGURE 4.22 This figure illustrates the effectiveness of comparison of model predictions with experimental data, which even simple on-line computer installations make possible in nuclear experimental work. Shown here are elastic scattering cross sections for ^{16}O ions on ^{16}O nuclei as functions of both energy and angle measured in the center-of-mass system. The upper figure shows actual experimental data obtained with an on-line computer-based data–acquisition system; the lower figure shows the corresponding model predictions. From examination of such comparisons of data surfaces rather than isolated angular distributions or excitation functions it is possible to determine at a glance whether the model correctly reproduces gross features of the experimental data without the confusion that otherwise arises from local detailed fluctuations.

[76]

gross properties of size and binding energies of nuclei and the chief
ingredients of the shell model—the central force and the residual interaction between the valence nucleons. This fundamental problem is not yet
wholly solved, for the agreement with observation is far from quantitative.
But this work is well begun.

Nuclear physics owes much in terms of insight, concepts, and techniques of calculation used in advancing the theory of nuclear structure
to molecular physics and the physics of condensed matter. The concepts of
molecular physics have been extremely important in treating nuclear
rotations and vibrations; the pairing concepts evolved to explain superconductivity in condensed matter also have played a crucial role in the
nuclear physicist's understanding of the excitation spectra of nuclei and
of the phenomenon of nuclear superconductivity. Much vital input concerning the nature of the nucleon–nucleon interaction has come, of course,
from elementary-particle physics.

The early history of nuclear-reaction research was dominated by the
phenomenon of the very narrow resonances discovered with low-energy
neutrons. Such resonances correspond to the formation of a configuration
of the nuclear system that lives a relatively long time before breaking up
with the emission of a nucleon, a photon, or a more complex nuclear
fragment. The postwar developments in experimental techniques made
possible neutron beams of much greater energy range. A new phenomenon
was uncovered, complementary in character to the nuclear resonance.
A broad energy structure, which varied gradually and systematically with
the atomic number of the bombarded nucleus, was found in neutron
scattering. These regularities could be well described in terms of the
scattering of neutrons by a smooth central force field of nuclear dimensions; in contrast to the longer-lived resonance phenomenon this is a
prompt mechanism. In addition to the scattering, of course, some absorption of the incident neutron beam occurred, and it was recognized that the
situation was analogous to the interaction of an incident light beam with
a rather cloudy crystal ball. This optical model clearly has connections
with the single-particle picture of the shell model. These concepts proved
applicable to all nuclear projectiles.

The prompt mechanism applies not only to simple scattering but also
to a great variety of transfer reactions in which one or a few nucleons
are transferred between projectile and target. Because the process is simple
and direct, these direct transfer reactions have provided a well-understood
means for studying the way nuclear excitations are built up, and very
much of the present detailed knowledge of nuclei comes about in just
this way.

In the 1960's, nuclear physicists addressed themselves to the discernment
and study of reaction cross-section structures intermediate between the

[77]

broad ones of the optical model and the very narrow resonances. The underlying nuclear configurations are understood to be of a complexity intermediate between the individual orbits of the optical potential and the complications of the narrow resonances, which in many cases have been interpreted as shell-model excitations. These intermediate structures have implications for the unraveling of the nuclear dynamics and for nuclear structure.

An example of such intermediate resonance phenomena is the isobaric analog-state excitations in heavy nuclei that have been discovered and extensively investigated in recent years. The isobaric spin quantum number, discovered in the early days of nuclear physics, was long thought to be useful only in light nuclei, where the Coulomb forces resulting from the proton charges were still too small to destroy significantly the charge symmetries that the isobaric spin quantum number described. Recently, with the advent of precision techniques applicable to heavy nuclei, it was found that this symmetry has much wider validity than was anticipated; in retrospect, the reasons are clear. The isobaric analog of a state is one in which structure is the same except that a neutron has been changed into a proton, and therefore its level is higher by the extra Coulomb potential energy. In heavier nuclei, this extra energy lifts the analog excitation into the continuum. The importance of this finding for further nuclear-structure studies is enormous. It shows that states whose structure is known from the low-energy work on their analogs are available at high enough energies for many reactions to be possible. The reactions to final states carry, then, a great deal of information about the nature of these final states. A new rich technique resulted that already has proved highly effective. Further, shell-model concepts apply with textbook simplicity in these newly accessible heavy nuclei—much more so than in the lighter regions where they have been studied previously. (See Figure 4.23.)

So far only the relatively simple excitations and reactions have been mentioned. But perhaps the most important characteristic of nuclear physics is the diversity of phenomena that manifest themselves. The fission of nuclei into massive pieces was discovered more than three decades ago, quickly applied in a national emergency, and, subsequently, stimulated entirely new industries; it was studied intensively throughout this entire period. Discovery of a new and very interesting fission phenomenon, the interpretation of which has great import for the field as a whole and especially for its frontier extensions, occurred during the late 1960's in the Soviet Union. It had long been thought that once a nucleus had been stimulated to fission, it did so very rapidly. The Soviet scientists found, surprisingly, that many nuclei had excited states with a very much longer lifetime before fission than did the normal ground states—the

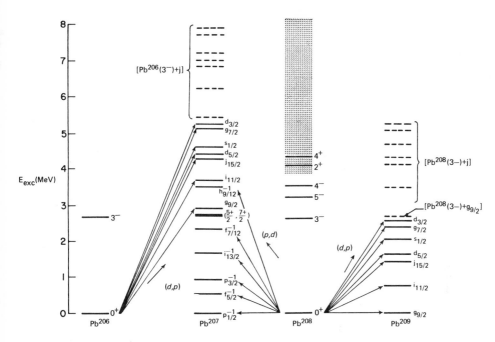

FIGURE 4.23 Experimental single-particle and single-hole levels in the lead region. ^{208}Pb is doubly magic, having 82 protons and 126 neutrons. ^{208}Pb states are formed by adding a single neutron to the ^{208}Pb core; ^{207}Pb states are formed by removing a single neutron from the ^{208}Pb core *or* by adding a single neutron to the ^{206}Pb core. As illustrated, all these states are very conveniently studied by deuteron stripping and pickup reactions. In addition to single-particle and single-hole states based on an unexcited ^{208}Pb core, it has become possible to study equivalent states based on this core in any of its excited configurations; shown here are the states that are based on the octupole vibrational 3$^-$ state at 2.6-MeV excitation in ^{208}Pb. While the ground state of ^{207}Pb is formed by the removal of a $p^{\frac{1}{2}}$ neutron from the ground state of ^{208}Pb, the closely spaced doublet (5/2$^+$ and 7/2$^+$) at 2.6 MeV of excitation is formed by removing a $p^{\frac{1}{2}}$ neutron from the 3$^-$ excited configuration of the core. Such systematic spectroscopic studies wherein the excited configurations are assembled, nucleon by nucleon, have provided extensive new nuclear structure information and have shown that the shell model concept of single-particle and single-hole orbits applies with classic simplicity in the lead region. Before the recent advent of the large electrostatic accelerators, this region of heavy nuclei was simply inaccessible to such high-precision studies.

so-called fission isomers. A massive international effort led to an understanding of these phenomena in terms of an outer balcony, which can develop in the nuclear potential barrier. The fission fragments, in essence, after penetrating the main barrier, can be trapped in the balcony, or

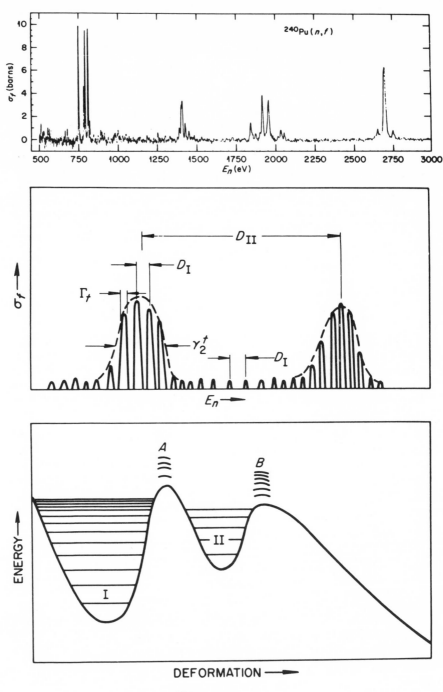

◄ FIGURE 4.24. The upper figure shows experimental data on the variation of the cross section for neutron-induced fission of plutonium-240 as a function of the neutron energy. Particularly striking are the periodic maxima in the cross section, with, in each case, a number of sharper resonances participating in the maximum. This phenomenon has led to important new insight into the mechanism of nuclear fission, as illustrated in the bottom figure, where the effective potential barrier is plotted against a parameter measuring the departure of the plutonium-241 compound nucleus from sphericity. In the past it was assumed that such a potential had only a single minimum (I) corresponding to the equilibrium shape of this nucleus and a single potential barrier through which the fission fragments tunneled before release. What the experimental data shown here have indicated, however, is that a second minimum (II) exists. How this reproduces the experimental results is shown schematically in the center figure. In some appropriate neutron energy range E_n, the compound, plutonium-241, is produced by neutron capture by plutonium-240 at a level of excitation such that the quantum states in the first potential well are separated by characteristic spacing D_I; because the second potential well is less deep, its quantum states at this same energy are more widely spaced (D_{II}). An enhanced correction occurs when the fission fragments tunneling through the first barrier find themselves at a quantum level energy in the second, where there is enhanced probability of tunneling through the outer barrier. These double potential shapes have now been recognized as being derivable from more microscopic nuclear shell models; they provide a textbook example of quantum-mechanical tunneling through barriers.

secondary minimum, for relatively long periods. Quite apart from some highly useful consequences, this finding constitutes a textbook example of penetration through a complex barrier. (See Figure 4.24.)

The existence of the second minimum had been predicted on the basis of shell-model calculations of the energy of nuclear deformation. That these calculations are closely related to those predicting the existence of as yet unseen islands of supertransuranic nuclei has provided no small part of the excitement (see Plate 4.III).

Future Developments

The frontiers of nuclear physics research are diverse. One of the main directions of general progress in the field has been toward the establishment of a coherent central model that reaches into all its phenomenological branches. As a result of this effort, the location of as yet uncharted regions has become clearer, and puzzles in the more definitely formulated areas have been identified. The future progress of nuclear physics will depend on a double-pronged attack against both kinds of problem.

Much of the progress in nuclear physics resulted from analyses of systematic studies of many types of systems and many nuclear properties. These studies must be extended with the newer techniques that have only recently become available. From 1900 to 1970, some 1600 nuclear species were identified and studied; reliable estimates suggest that collisions of 2-GeV uranium nuclei with uranium targets will produce at least 6000 different species. In addition, there is relatively little information as yet concerning the behavior of any nucleus as more excitation energy is pumped into it. Very recent evidence suggests that, in addition to heating and evaporating nucleons or fragments in analogy to a water drop, quasi-molecular complexes of long lifetime may form in which most of the available energy is bound in molecular fragments with the remaining available energy appearing as kinetic energy of simple relative motion of these fragments. This high-excitation energy region is still largely *terra incognita*.

Although much of this work can be done with existing techniques and probes, a quite different set of probes rapidly is coming into increasing use. New accelerator sources and detection devices now under construction will lead to precise nuclear investigations by high-energy protons and extend the capabilities of high-energy electrons. Examining the nucleus with these two different short-wavelength beams will correspond to putting the nucleus under high-resolution microscopes illuminated by complementary radiations. New copious sources of mesons will be used to probe the nucleus for different specific components of nuclear motions (see Figure 4.25). Mesonic atoms, formed by mesons orbiting the nucleus, sampling and reporting on the nuclear matter that they traverse, will provide still different complementary nuclear information. The intensive study of hypernuclei, formed by replacing one of the constituent neutrons or protons by a strange particle—a hyperon uninhibited by the Pauli Exclusion Principle—will provide yet another view from deep inside the nucleus. The possibility of the existence of a region of stable superheavy nuclei, lying well beyond the heaviest now known, has been much discussed. If current ideas on element production are correct, the superheavies are beyond nature's capabilities and can be created only by a leap over the instabilities that surround them. This leap can be achieved by the bombardment of massive nuclei with massive projectiles at energies sufficiently high for them to overcome their mutual repulsion and fuse. Such reactions appear also to be fruitful sources of the many other nuclear species that are sought. The new heavy-ion facilities now being planned will begin to open these possibilities to exploration with as yet unknown nuclei. If the superheavies do indeed exist, current calculations suggest that, among other things, they should emit about three times as many neutrons per fission as do the present fissile fuels. This capability could have major consequences in the development of a more convenient, perhaps portable, energy source.

Superheavy Nuclei

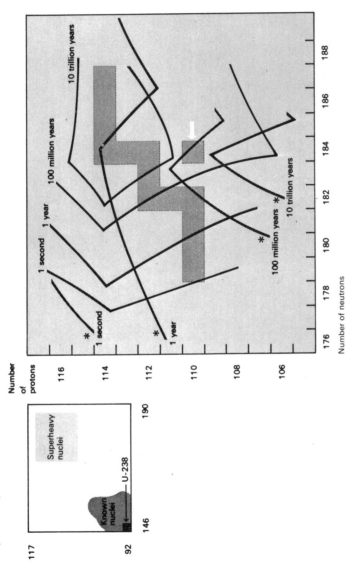

PLATE 4.III Superheavy nuclei. Theoretical calculations predict great stability for several superheavy nuclei. Starred lines show stability against alpha-particle decay, unstarred lines against spontaneous fission, and solid squares against beta-particle decay. The most stable nucleus would thus be one having 110 protons (arrow) and several closed-shell nuclei having 114 protons. [Source: *Science Year. The World Book Science Annual*. Copyright © 1969, Field Enterprises Educational Corporation.]

FIGURE 4.25 Aerial view of the Los Alamos Meson Physics Facility. This facility features 1-mA proton beam at 800 MeV, plus simultaneously either 100-μA hydrogen ions at 800 MeV or 1-μA polarized hydrogen ions at 800 MeV. Beams of protons, neutrons, muons, and neutrinos will be available for simultaneous use in a multidisciplinary program.

The use of heavy projectile beams is the most rapidly developing frontier in nuclear physics both here and abroad. These beams have become available under adequately controlled and precise conditions only in recent years and already have been widely exploited. They not only permit exploration of known nuclear phenomena in ranges of parameters —energy, angular momentum, and the like—well beyond any previously available, but they also give access to totally new phenomena—totally new configurations (the nuclear molecules and possibly superheavies are examples) and totally new dynamics (How do two heavy nuclei deform during a collision? What happens to this deformation energy?) Massive construction programs directed toward development of forefront facilities for such investigations are currently in progress in the Soviet Union, Germany, and elsewhere.

Applications of Nuclear Physics

The discovery and use of nuclear radioactivity, dating back to 1913, has provided research and clinical workers in the biological and medical sciences with a probe of unprecedented specificity and power and has revolutionized these sciences; nuclear radioactivity has provided a technological treasure trove that, even now, remains only little exploited. The discovery of nuclear fission in 1938 fundamentally changed the nature of both peace and war. At the same time, and much more important, it provided mankind with a wholly new resource of nuclear energy. In loosing the constraints imposed by Nature's caprice in locating her fast-dwindling energy resources and by civilization's prodigal use of them, nuclear energy stands as one of man's major weapons in his continuing struggle against poverty, hunger, and despair.

Less well known than radioactivity and nuclear fission but, in sum, perhaps equally important are the myriad small inventions, applications, and ideas that have been a part of nuclear physics since its inception. Even a cursory inspection of the intricate instrumentation in the intensive-care section of a modern hospital, the control section of a modern manufacturing plant, the nerve center of a major defense establishment, or the control rooms of Cape Kennedy or Houston reveals the debt owed to those pioneers in nuclear physics, who, often by intuition as much as logic, invented and devised the instruments that gave them a glimpse of the nuclear world. The commercial application of nuclear radiations is in its infancy but expanding rapidly; it ranges from the giant irradiation units performing vital but unseen service on production lines to the minuscule sources energizing, in virtually eternal and foolproof fashion, the emergency and warning signs now used around the globe in languages from Hindi to English. Medical applications of nuclear radiations run the gamut from the now familiar clinical use of diagnostic x rays and of these and harder radiations in the treatment of carcinomas to the newer uses of much more sophisticated and specific diagnostic probes such as the Anger camera (Figure 4.26), which probes lesions in the depths of the brain; neutron radiography units that make visible for the first time the soft tissues deep in the body; and new isotopes whose use permits unraveling of ever finer details of the intricate biochemical and biophysical bases of life.

At the present time, taking the world population at 3.2 billion and the total yearly consumption as approximately 3.5 million megawatt-years, the average yearly energy consumption per person corresponds to about one ton of coal. Harrison Brown et al.* have estimated that by the year

* H. Brown, J. Bonner, and J. Wier, *The Next Hundred Years* (The Viking Press, New York, 1957).

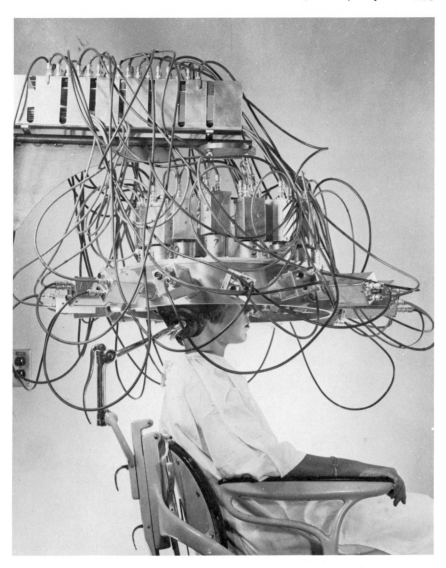

FIGURE 4.26 Brain tumor locator. The Anger camera forms an image of a portion of the brain. Gamma rays from radioisotopes pass through collimators to scintillators. Photomultiplier tubes detect signals, which are processed and displayed by a
computer. About 10 million scans on patients are performed each year (three fourths
of these in the United States). [Source: Brookhaven National Laboratory.]

2060, with a world population of 7 billion, this average equivalent per
capita energy consumption will have increased to ten tons—or to a total of

70 billion tons of coal per year. At this rate, and, if anything, Brown's estimates seem conservative, for desalting of seawater is not taken into account, the present fossil fuel reserves of perhaps 2400 billion tons of coal equivalent would last only some 50 years. This provides the dimension of the problem that faces contemporary civilization; Table 4.3 gives the projected energy production pattern developed by Brown *et al.* for the year 2060. These estimates assume that nuclear sources will supply some 65 percent of the total world energy requirements. If desalting of seawater is included, this fraction would rise to about 75 percent.

Similar estimates project that by the year 2000 nuclear energy will be furnishing one half of the total electrical energy of the United States. Annual investments in nuclear power plants during the next ten years are expected to average some $3 billion. Nuclear power is fast becoming a major industry, and a major program directed toward the development of breeder systems has only recently been announced by the federal government. This action is in recognition of the fact that available resources of uranium-235 are limited; current reactors extract only 1–2 percent of the energy potentially available from their uranium fuel and are dependent on

TABLE 4.3   Projected Energy Production Pattern for the Year 2060 [a] (World Population 7×10^9)

Source	Equivalent Metric Tons of Coal (Billions)	Equivalent Heat Energy	
		10^{18} Btu [b]	MW-yr Heat [c]
Solar energy (2/3 of total space heating)	15.6	0.42	140×10^5
Hydroelectricity	4.2	0.10	38×10^5
Wood for lumber and paper	2.7	0.07	24×10^5
Wood for conversion to liquid fuels and chemicals	2.3	0.06	21×10^5
Liquid fuels and petrochemicals produced via nuclear energy	10.0	0.27	90×10^5
Nuclear electricity	35.2	0.96	320×10^5
TOTALS	70.0	1.88	633×10^5

[a] From H. Brown, J. Bonner, and J. Wier, *The Next Hundred Years* (The Viking Press, New York, 1957).
[b] Btu: British thermal unit = 252 calories.
[c] MW-yr: megawatt-year.

the availability of large amounts of low-price ores to keep their power costs economically competitive with those of fossil-fuel plants. Breeder-reactor power costs are expected to be relatively insensitive to the cost of uranium and thorium, thus making possible economic use of low-grade ores. Development of a successful, commercially competitive breeder may well be one of the most important scientific and technological tasks facing mankind at the present time.

That nuclear data for these systems is not yet known with adequate precision frequently is forgotten or ignored. Barschall, one of the pioneers in the precise measurement of neutron cross sections, has noted that some of the most basic cross sections used in the design of slow fission reactors are uncertain by as much as 10 percent; in the case of the cross sections needed in breeder-reactor work at higher neutron energies, uncertainties in excess of 50 percent are common. The nuclear engineer, faced with these uncertainties, is forced to more conservative and more costly designs to incorporate larger margins of safety. The absence of these data simply indicates that the necessary measurements are both difficult and sophisticated; only very recently have they become possible. For example, measurements on the ratio of the neutron capture to fission cross sections for ^{239}Pu, when carried to presently available precision, demonstrated that one particular breeder-reactor system, on which over $100 million in development capital was planned, was not feasible.

In the case of conventional reactors based on the use of thermal neutrons, the designs are not seriously limited by uncertainties or unknowns in the nuclear data. But even these limits involve very high stakes. Among the uncertainties are those associated with nuclear data affecting primarily the fuel costs. These uncertainties are estimated to be of the order of only 0.02–0.03 mil per kilowatt-hour of electric power. But applied to the entire projected nuclear power capacity of the United States, even this small uncertainty will amount to $20 million to $30 million in annual fuel costs by 1980 and $140 million by the end of the century. And in the case of breeder reactors, the uncertainties and possible cost savings, at present, are much larger.

Further off on the nuclear power horizon is the fusion reactor (see section on Plasmas and Fluids). While the main effort is in plasma physics, nuclear physics and physicists are involved in a number of essential ways with the primary nuclear phenomena. Also on the far horizon is an accelerator-based power system, contemplated by Lawrence in 1948, then shown to be much ahead of its time technologically but recently reinvestigated extensively by Canadian groups. This system takes advantage of the fact that neutrons produced by collisions of 1-GeV protons with matter cost roughly ten times less in energy than do those produced in fission; since neutron economy is all-important in fission energy sources, this low-cost aspect is most attractive. In principle, this system appears less attractive than fusion, but it is necessary to maintain a level of activity in related research and technology such that it would be possible to move forward if major breakthroughs here—or major disappointments with fusion systems—should so dictate.

The many contributions of nuclear physics to the medical, industrial, agricultural, and general technological fields are described in detail in the report of the Panel on Nuclear Physics. Especially in technological fields, a close and fruitful symbiosis has evolved. Nuclear physics interacts with all these fields through five primary channels: radioisotopes as tracers, radioactive nuclei as energy and radiation sources, nuclear methods of materials analysis, direct utilization of electron and ion beams from accelerators, and, finally, the great diversity of instruments developed by nuclear physicists, or by others, in response to particular needs of nuclear research.

The availability of radioisotopes of all elements—and when desired, in large quantities—has worked revolutionary changes in many fields but perhaps most strikingly in medicine and the biological sciences. The specific example of the isotope ^{99}Tc is instructive. Technetium ($Z=43$) was early recognized as one of the two elements missing from the Mendeleev chart of the elements as far as terrestrial abundance was concerned. Identified in the 1930's as responsible for certain unassigned lines in the optical spectra of some unusual stars, technetium was first recovered in measurable quantities in the 1940's from an old and much-used molybdenum septum from the Illinois cyclotron, where it had been produced by the (p, n) reaction. Still it remained a laboratory curiosity.

By 1964 it had been recognized that ^{99}Tc had unique qualifications for use in brain scans. It had low-energy characteristic radiations that permitted its detection and identification with minimal extraneous or unnecessary patient exposure; it concentrated selectively and rapidly in diseased tissues; its radiations were readily collimated, permitting precise location of the isotope concentration; and its half-life (6 h) rapidly removed it from the body. In 1969, a substantial fraction of the entire Oak Ridge isotope separation effort was devoted to an attempt to meet the urgent demands of physicians throughout the country, and it has been estimated that ^{99}Tc production (through accelerator bombardment) and distribution will shortly form the basis for a multimillion-dollar-per-year industrial operation. Little did the nuclear scientists intent on separating micrograms of technetium from the defunct Illinois septum in the 1940's realize that kilograms of the element would be the production unit in the 1970's, or that in 1969 alone use of their element would be responsible for fending off death in literally thousands of cases. This example is perhaps extreme, but it illustrates the tangled and unpredictable linkage between discovery and ultimate use.

The growth of radioisotope utilization in industry has been phenomenal. In 1969, about half of the 500 largest manufacturing concerns used radioisotopes. About 4500 other firms also are licensed to use radio-isotopes. Virtually every type of industry is represented. The growth continues as new radioisotopes become available and as detector and instrumentation improvements continue to increase the sensitivity, selectivity, and reliability of these techniques. The estimated saving to U.S. industry resulting from the use of such methods in a myriad of flow-rate applications was $30 million to $50 million in 1963—a year for which statistics are available—and has increased greatly since then. Estimates in 1969 indicated that radioisotope gauges, together with their associated instrumentation, comprised a $35-million-per-year market that was expanding rapidly. The cost of nuclear oil-logging techniques amounts to some $25 million; the savings to the petroleum industry are many times that. Sterilization of medical supplies and materials by nuclear radiations has been put into routine production use and has led to new industrial ventures.

The technique of neutron radiography, long a workhorse in the field of nondestructive testing (see Figure 4.27) has been greatly enhanced by the availability of the new transuranic neutron-emitting radioisotope, californium-252.

Accelerators, developed as part of the nuclear-research program, have been applied to many other purposes. Over 1000 such machines are at work in medical, industrial, and technological operations. Some 200 accelerators around the country are furnishing the radiation therapy required by more than 300,000 patients yearly. New accelerator developments are quickly translated into improved radiation facilities. Thus, the high-energy machines, developed in the nuclear program, offer a narrow pencil of radiation that allows the delivery of more radiation to the lesion and less to surrounding healthy tissue. New advances in the technology of linear accelerators were quickly recognized as offering more efficient radiotherapy devices; many units are already in service, while others are being installed. Industry has built accelerators into many other functions: radiation processing, industrial radiography, and neutron radiography among others. The investment in such accelerators, over $130 million, affords some measure of these applications, independent of instrumentation or plant costs. Radiation-processed products manufactured annually amount to over $1 billion in sales.

The instrumentation of nuclear physics plays an important part in all the above applications and in many other ways. The scientific instrumentation industry is a key one, providing the tools on which much tech-

[91]

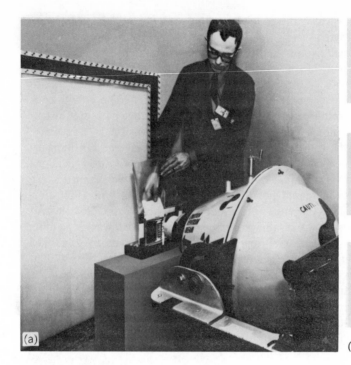

photograph

x-ray radiograph

(a)

(b) neutron radiograph

| Photograph | Neutron Radiograph | X-Radiograph |

[92]

◄ FIGURE 4.27   *Top:* Nondestructive testing. Neutron radiography. (a) Recently developed neutron-radiographic camera, based on californium-252, offers the significant advantage of portability (100-lb unit), so that the exposure may be conducted in a manufacturer's plant or at a field service facility, thus avoiding the necessity of shipping the item being neutron-radiographed to an atomic reactor facility. (b) Comparison of photograph, x-ray radiograph, and neutron radiograph of bullets. (Battelle Northwest.) [Source: John R. Zurbrick, *Yearbook of Science and Technology, 1971.* Copyright © 1971, McGraw-Hill Book Company, Inc. (Used with permission of McGraw-Hill Book Company.)]

Bottom: Comparative radiographs of a combination lock. [Source: This photograph was provided through the courtesy of Atomics International, a Division of North American Rockwell Corporation.]

nological progress depends. The nuclear-instrument section of this industry is large and rapidly growing beyond the $100 million sales volume estimated in 1970. Nuclear instrumentation sales abroad are also significant; in addition, they unite U.S. technology with that of the rest of the world.

Nuclear physics affects our technological society in other less tangible ways than those exemplified by its tools and products. The development of the computer industry provides an illustration. Much of the original logical circuitry was based on the electronics developed for nuclear-data purposes. Nuclear physicists were among the first to use a computer not only to analyze raw data instantly but also, acting on this analysis, to control an experiment. Such on-line processing is now common throughout industry, in high technology as well as the routine production of basic commodities. The sophisticated demands of nuclear applications have in themselves provided important stimuli for computer developments. Thus the need for analysis of nuclear explosions stimulated the development of the very large digital computer and its associated software. At the other extreme, the rapidly expanding minicomputer market owes part of its success to the pioneering use of these small processors in nuclear installations.

Much of the world's military arsenal is nuclear. Almost all of physical science is involved in describing a nuclear explosion or its effects; however, nuclear physics is central to this effort. Nuclear physics is equally relevant to establishing and then monitoring international test ban and disarmament agreements, the intention of which is to lessen or eliminate the chance that nuclear explosives will ever again be used in war. Nuclear physics and physicists continue to play important roles in every aspect of the U.S. national defense effort: the design of nuclear weapons, the testing program, the evaluation of weapons effects and a civil defense

[93]

program, test ban monitoring and surveillance, and the arms-limitation efforts.

Interactions with Other Sciences

Nuclear physics is an integral part of the physical sciences. The mesonic nature of the force between the individual constituents of nuclei forms a close connection with particle physics. Many basic phenomena first seen in the nuclear domain apply in all of physics and the nucleus provides a well-explored laboratory. The weak interactions were first seen in nuclear radiations, and the violation of the fundamental parity law was demonstrated in a nuclear experiment. The concepts evolved in the course of developing nuclear physics have become part of the milieu of physical ideas in which physics as a whole grows.

Astrophysics is perhaps most closely and particularly related to nuclear physics, since nuclear processes produce much of the energy and achieve the elementary composition of the matter in the universe. A special subfield, nuclear astrophysics, concentrates on these problems. Both the problem of stellar evolution and the nucleosynthesis of the elements depend on knowing the dynamic features of reactions among a great variety of light and heavy nuclei at the energies of stellar processes. Although study of these problems constitutes a separate subfield, it is one that remains closely linked to the progress of nuclear physics. Thus the recent experiment on the measurement of the flux of solar neutrinos by the inverse of nuclear beta-decay has sharply thrown into question the picture of solar energy production that has been developed over the years. The exciting conjectures on the existence of neutron stars and other manifestations of superdense agglomerations of matter offer another and rather different example. It is a long extrapolation indeed from several hundred nucleons to 10^{56} nucleons, from normal nuclear densities to those so great that strange particles must be considered along with the nucleons, and from nuclear masses to those so enormous that nuclear and gravitational energies vie with one another. This domain is wholly new, one that tests the ideas and methods of nuclear physics, but one for which a solid conceptual foundation has been laid.

Solid-state physics and nuclear physics have interchanged ideas and tools for many years. Concepts based on many-body theory have proven of fundamental value in explaining nuclear phenomena; thus the formalism developed in understanding superconductivity has explained the pairing correlations in nuclear wavefunctions. The theory developed for nuclear matter has served for the many-body problems of condensed matter. The use of radioactive nuclei to probe their electric and magnetic environments in solids has been put to use in many problems of solid matter. The

Mössbauer effect is now a tool of the solid-state physicist. The use of neutrons to study the static and dynamic characteristics of solids is so well established that such work pre-empts the largest share of research time on the most modern reactors.

Chemistry, too, has formed two-way bonds with nuclear physics. A substantial fraction of the nation's nuclear research effort takes place in chemistry departments and laboratories. Hot-atom chemistry forms a classical region of overlap. Very-heavy-element chemistry and the discovery of new very heavy elements have always been directly related. Chemistry, like solid-state physics, has profited from the neutron capabilities of research reactors. An interesting new effort of chemical kinetics and structure is based on the study of elementary atom–atom, atom–molecule collisions, thus drawing directly on the methods and concepts of nuclear-reaction work—just as earlier nuclear physics took over the developments of atomic scattering methods.

From the time of the earliest studies of radioactivity, scientists realized that there was at hand a dating method for geological processes, and, by 1905, the first dating of rocks had been accomplished. Geochronology is now a fundamental part of the earth sciences and employs a great number of isotopes and isotope chains in the study of the time domain from one million to five billion years. Work on meteorites, lunar samples, and ocean-bottom-sediment cores is of wide public interest. Archeological studies based on the cosmic-ray-induced activity in carbon have provided a time scale commensurate with man's recent historical period. Radiocarbon dating is now a standard working method of the archeological laboratory.

Art history and forensic investigations also have profited from neutron activation analyses. Beyond this, nuclear systems provide the only effective clock, which has been running continuously since the formation of the solar system, and thus provide the possibility of testing cosmological speculations such as that concerning the fundamental physical constants and their possible variation with time over cosmological periods. The possibility that the elementary charge unit (that carried by the electron and proton) might vary in such a way that its square was proportional to time had been discussed for decades. Recently, examination of the stability systematics of heavy nuclei has demonstrated conclusively that, if this charge varies at all, its variation is less than 0.3 percent of that postulated by the cosmological arguments.

The biological and medical sciences have used the whole battery of radioactive isotopes to trace life processes. The materials as well as the detectors and instrumentation to measure their progress in biochemical reactions were the products of the nuclear laboratory.

Only recently, as part of the MAN Project at Oak Ridge, Anderson

[95]

and his collaborators have begun to use centrifuges, developed in wartime for possible use in the separation of uranium isotopes, for biological research. For example, there have always been significant numbers of persons who were unable to receive protection from influenza because of marked allergy to the protein contaminants in the influenza vaccine. This inability to receive such protection was particularly serious in the case of older persons. During the past year alone over 50,000 of them were able to receive protection, which would otherwise have been inaccessible to them, because Anderson and his colleagues had discovered that ultra-centrifuging the influenza vaccine effectively removes all the protein contaminants. This work is only the beginning of a major instrumental change in the production and purification of biologicals.

Nuclear physics has also had a continuing interaction with the space-science effort. Nuclear instrumentation was a crucial part of space probes to measure the fluxes of high-energy electrons, photons, and nuclear fragments that form important parts of the environment of space. The radiation effects on spacecraft and the space traveler have been studied with nuclear accelerators. The first on-site studies of the lunar material were made by a nuclear scattering device; the instrumentation left behind to record lunar events is powered by nuclear isotope sources. These are, no doubt, mere preludes to man's future probing of more distant planets. The nuclear tools will grow in importance. Nuclear-powered rockets, already developed in prototype, may well be the essential means to move the huge payloads that currently are necessary. The connection will go much deeper than devices and instruments, for space science and nuclear physics share the problem of analyzing and understanding cosmic rays.

The Organizational Structure of Nuclear Physics

Nuclear-physics research is divided about equally between universities and government laboratories. A little over half of the scientific effort is university-based; most of the rest takes place in the AEC National Laboratories. If one includes both federal and nonfederal support, then the funds for operating these facilities are also about equally divided.

The essential nature of nuclear physics requires a broad effort with many kinds of programs. The present program utilizes more than 100 facilities—potential drop machines, cyclotrons (see Figure 4.28), and linear electron accelerators—ranging from those capable of being used by two or three scientists to major installations that require a large, trained crew to operate. The number of separate projects based on these facilities, or not requiring machines, is many times larger. There is a nearly continuous distribution of university-based projects from the very small to those of the same order as the large National Laboratory programs.

[96]

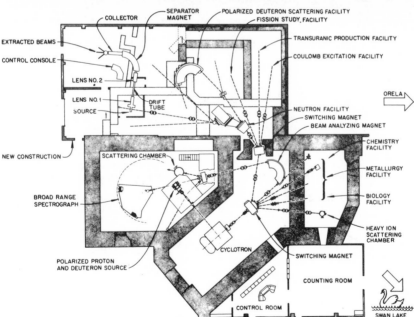

FIGURE 4.28 *Top:* The beam lines in the main cyclotron vault. The rf system of the cyclotron appears at the right and the beam preparation magnet at the left.

Bottom: A layout of the beam lines of the Oak Ridge Isochronous Cyclotron, showing the bending and analyzing magnets and the various shielded research rooms.

[Source: Oak Ridge National Laboratory.]

[97]

NUCLEAR PHYSICS

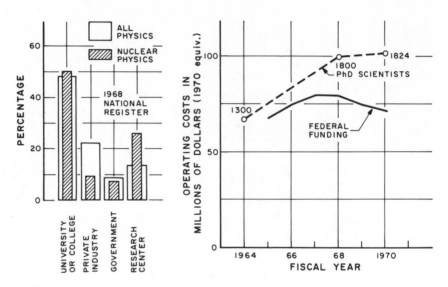

FIGURE 4.29   Manpower, funding, and employment data on nuclear physics, 1964–1970.

NUCLEAR PHYSICS

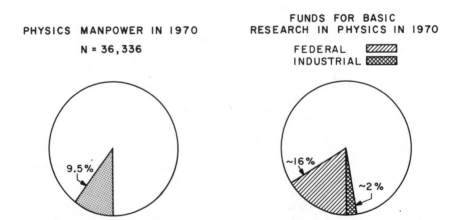

FIGURE 4.30   Manpower and funding in nuclear physics in 1970.

[98]

Nuclear physics has been largely an experimental program; only 15 percent of the scientific effort is devoted to theoretical research. The main experimental efforts, comprising about half of the total experimental program, have been based on Van de Graaff and cyclotron accelerators. This emphasis, of course, will shift as the facilities designed to open new phenomena to investigation come into operation, but a broad effort is still to be expected.

Nuclear physicists comprise about 13 percent of the physics PhD population, as reported in the 1970 National Register of Scientific and Technical Personnel. They are generally academically based; however, the research center is a close competitor for people. The number of PhD's working in the subfield grew at a rate of about 10 percent per year to 1968 but has since leveled off (Figure 4.29). In 1970, approximately 16 percent of federal funds and 2 percent of industrial funds for basic research in the physics subfields were allocated to this subfield. The federal entry includes $12 million in funding for basic nuclear-physics research supported under chemistry (Figure 4.30).

Problems in the Subfield

Funding trends in nuclear physics since 1967, if continued beyond the present, would appear to imply a decision to phase down activity in the subfield to a level at which it could no longer hope to remain at or near the frontier in major areas of current activity.

Three major areas must be considered. First is the broadly based activity involving a wide variety of techniques, facilities, and approaches that constitutes the central core of nuclear physics and the base on which all future activity in the science and its external utility must rest. Second is the group of major new facilities that received approval in the mid-1960's and are only now coming on-line, with corresponding large demands on dwindling operational funds; the Los Alamos Meson Physics Facility is the most visible example. Third are the major new starts that must be undertaken in the subfield in the next few years if it is not to stagnate or withdraw from the developing frontiers; a national heavy-ion physics facility is the outstanding current example. No one of these areas is viable over an extended period of time without the other two.

And as regards the Atomists, it is not only clear
what their explanation is; it is also obvious that
it follows with tolerable consistency from the
assumptions which they employ.

ARISTOTLE (384–322 B.C.)
On Generation and Corruption, Book I, 325

ATOMIC, MOLECULAR, AND ELECTRON PHYSICS

Introduction

At the beginning of the twentieth century it became clear that the atom was not indivisible but consisted of a small, heavy nucleus and a cloud of orbiting electrons. Although deeper probing of the nuclear structure led to the development of nuclear and high-energy physics, the detailed study of the properties of the electron clouds in atoms and molecules remained the subject matter of atomic, molecular, and electron physics.

It should be emphasized that these properties determine the structure of all chemical and biological compounds, and the forces between the electrons are also responsible for the cohesion of matter in liquids and solids. In fact, the electrical interactions between electrons and nuclei in atoms and molecules largely determine the physical phenomena of everyday life, for example, the emission of light in fluorescent bulbs, the boiling temperature of water, the course of electron beams and the lighting of a screen in television tubes, and the nature of chemical reactions such as the burning of coal or oil.

Clearly, atomic, molecular, and electron physics occupies a central position among the sciences and in the science curriculum. This subfield reached perhaps its apex as a research frontier in the first quarter of this century. The spectroscopic investigations of light emitted by atoms and molecules established the facts on which the central physical theory of quantum mechanics is based. In the late 1920's and early 1930's, the understanding of chemistry made great strides based on the quantum theory of electron orbits. Electronic technology was also rapidly developed during this same interval and spurred the revolution in communications that is exemplified by radio and television.

With the basic principles so well established, many physicists considered research in this subfield essentially complete; the most concentrated effort in physics since 1930 has been directed toward nuclear and high-energy physics. However, atomic, molecular, and electron physics continues to contribute substantially to the understanding of natural phenomena. Four Nobel Prizes in each of the past two decades were awarded in this subfield. The interaction of electromagnetic fields and electrons has been pushed to new orders of magnitude in precision and intensity. Within one decade of its invention, originating in atomic and molecular physics, the laser has become a household word. Applications of this instrument appear not only in the physics laboratory but also in hospital operating rooms, high-quality machine tool lathes, military range finders— and even in James Bond movies!

Important Recent Developments and Current Activity

During the past ten years, the development of lasers and masers stands out as a critically important scientific achievement with broad technological relevance. The atomic hydrogen maser is the most accurate clock devised by man and makes possible timekeeping with an accuracy of 1 sec in 30,000 years. Improvements in navigational systems are an obvious by-product of this development. Lasers constitute a radically different and new type of light source, characterized by the extreme directionality, intensity, and color definition (wavelength) of the emitted light (Plate 4.IV). This new light source is used to measure distances with unprecedented accuracy for such diverse purposes as the precise control of machine-tool operations (Plate 4.V), the measurement of distances between points on earth (geodesy, continental drift, and crustal motion preceding earthquakes), and the measurement of the distance to moving objects (light radar, including ranging of low-flying aircraft and the distance to the moon). Measurement of the distance to the moon was accomplished with an uncertainty of only 6 in., which gives an idea of the high level of accuracy and great potential of this instrument.

The laser has rejuvenated the centuries-old science of optics and has created the entirely new subfield of nonlinear optics, the study of the necessary modifications of the well-known optical laws of refraction and reflection at extremely high light intensities. A new branch of technology, optoelectronics, is developing. The newly acquired physical knowledge of lasers and nonlinear optics is applied to optical communications, in which signals are transmitted and processed by light beams. Three-dimensional optical displays, the picturephone, and optical computer elements probably will be widely applied during the coming decade.

Another highly active field of investigation is the study of the interaction of individual molecules in specified states of vibration and rotation. The experimental technique makes use of molecular beams colliding in vacuum and gives much more detailed information about the causes and fundamental characteristics of basic chemical reactions than was previously attainable. This focus on the elementary chemical interactions holds high promise of an entirely new understanding of chemical processes. Use of merging beams, in which the relative velocities can be adjusted to permit

[101]

study of collisions at relative energies measured in small fractions of electron volts, opens still further new vistas in basic collision phenomena.

Electron beams in high vacua are also used in electron microscopy, in which under favorable circumstances individual atoms have recently been made visible (see Figure 4.31), and in microprobe analysis of impurities and surfaces. The improved vacuum and beam-handling techniques used here hold high promise of increasing the present inadequate understanding of the physics of surfaces. The burgeoning low-energy electron

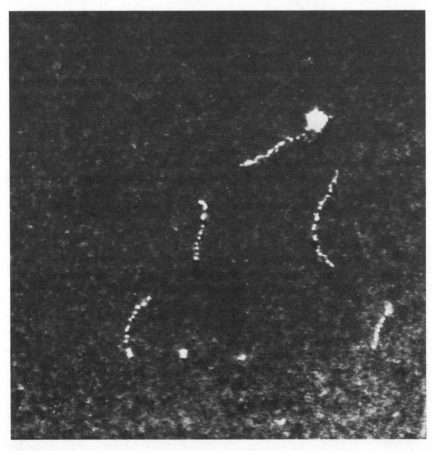

FIGURE 4.31 Single atoms and multiples of single atoms of thorium in the polymeric organic salt of 1,2,4,5-benzene tetracarboxylic acid are photographs on a 25-Å thick evaporated carbon film. The photograph is roughly 2100 Å to a side and was taken with the high-resolution scanning electron microscope in the laboratory of A. V. Crewe.

PLATE 4.IV Mode pattern of coherent light from an optical gas maser. To produce these mode patterns the normal operation of a helium–neon optical maser is
perturbed by placing a pair of wire cross hairs in the cavity. These wires interact with
the mode structure of the unperturbed cavity, suppressing some modes and, in certain
cases, coupling others together. By changing the angle between the cross hairs this
interaction can be altered and different mode patterns can be produced. [Source:
Bell Telephone Laboratories.]

PLATE 4.V   Experiments in laser machining techniques. [Source: General Electric Company.]

diffraction (LEED) field has opened up for study whole areas of surface physics for the first time.

The new field of beam-foil spectroscopy has rejuvenated the study of atomic spectra. In this work, high-velocity ions from accelerators are passed through solid foils in which very highly stripped and highly excited ionic species are prepared for spectroscopic examination in subsequent flight. Geometric shifting of the observation point permits detailed study of de-excitation and relaxation processes that previously were inaccessible. By this method higher states of ionization can be obtained in the laboratory than by any other method, and some transitions previously observed only in solar spectra have already been studied.*

Interaction with Other Subfields of Physics and Other Sciences

The nature of the collisions among electrons, atoms, and molecules has implications for a variety of other subfields and sciences. In fact, much atomic physics is supported and carried out in connection with other subfields, such as plasma physics, atmospheric physics, optics, and space physics. These elementary collision processes, which can now be studied in fine detail by colliding beam techniques, play a role in practical problems such as the re-entry of missiles and space vehicles into the atmosphere and the initiation and containment of plasmas necessary for controlled thermonuclear fusion; they are also extremely important in many problems in astrophysics, radioastronomy, chemistry, and biochemistry. (See Figure 4.32.)

Since the interactions between electrons and atoms determine most of the physical phenomena that man encounters, the central position of atomic and molecular physics is obvious. It contributes heavily to other disciplines and, in turn, benefits greatly from new developments in these other disciplines. For example, the development of microwave radar techniques during World War II contributed greatly to the rise of microwave molecular spectroscopy. New laser technology, born from atomic physics, provides better tools for spectroscopy and plasma diagnostics and has made Raman spectroscopy a practical method in chemical analysis.

The interaction with the physics of condensed matter is particularly strong, and the special field of quantum optics can be considered as a part of either subfield or of both. Laser phenomena in solids offer entirely new device possibilities, particularly when combined with some of the very

---

*Ad Hoc Panel on New Uses for Low-Energy Accelerators, NRC Committee on Nuclear Science, *New Uses for Low-Energy Accelerators* (National Academy of Sciences, Washington, D.C., 1968).

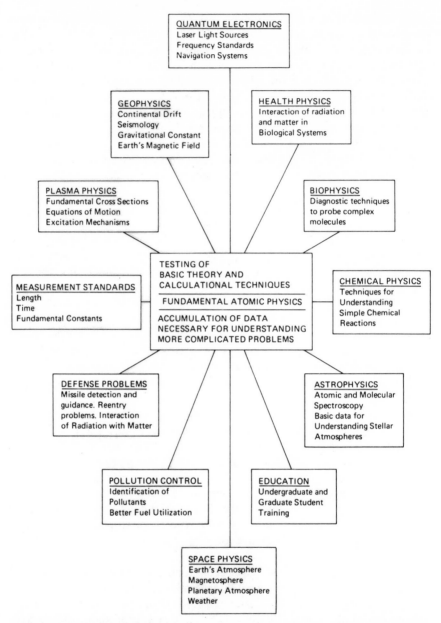

FIGURE 4.32   Relationship of atomic and molecular physics to other disciplines. [Source: Committee on Atomic and Molecular Physics, National Research Council, *Atomic and Molecular Physics* (National Academy of Sciences, Washington, D.C., 1971).]

recent tunable metal oxide surface situations developed in condensed-matter studies. LEED and other studies on clean surfaces provide not only the opportunity to study atomic and molecular interactions in two dimensions in monoatomic or monomolecular layers but also access to entirely new surface phenomena crucial to such applications as catalysis. With the rapidly increasing interest in the control of industrial pollutants, understanding of such catalysis phenomena holds promise of widespread and critically important societal applications.

Cold-field electron emission and field-ion microscopy have been developed to a point at which the diffusion of individual atoms along the surface and the influence of atomic configuration along different crystallographic planes can be followed visually. Spectroscopic relaxation techniques also are useful in the study of solid-surface interactions of atoms. Electron and ion-beam sputtering and ion implantation, and especially scanning-beam electron spectroscopy, provide other examples of the interaction of atomic, molecular, and electron physics with condensed-matter surface physics and high-vacuum technology. Another interface with the physics of condensed matter is the study of molecular fluids. Laser-beam light-scattering techniques are contributing very much more precise information about the mechanics and molecular arrangements in liquid crystal transitions and in mixtures of fluids. This is a research area on which atomic, molecular, and electron physics; condensed-matter physics; chemical physics; and biophysics all impinge.

Fundamental measurements of characteristically high precision (parts per million) in atomic and molecular physics have complemented those in elementary-particle physics in exploring the possible limits of validity of quantum electrodynamics. Such measurements provide crucial inputs to the precision determination of some of the fundamental physical constants and have other more esoteric physical uses. Atomic-beam measurements have been used to establish limits on possible mass anisotropies in the universe and to set stringent upper limits on possible differences in the electrical charge of various elementary particles. Atomic and molecular physics continues to stand at the forefront of ultrahigh-precision physical measurements. It has exported metrology and other precision techniques to all of science and engineering.

This subfield interacts also with nuclear and high-energy physics, as charged fundamental particles, such as positrons and mesons, antiprotons and hyperons, lead to the formation of artificial atoms, called, for example, positronium, muonium, and mu-mesic and pi-mesic atoms, in which one of the named particles substitutes for a normal atomic electron. Somewhat unexpectedly, the study of positronium annihilation in gaseous and condensed media paved the way for useful advances in chemistry and condensed-matter physics. The interaction of a polarized laser beam with an electron beam to obtain polarized high-energy gamma rays is another example of interaction with high-energy physics. Optical and radio-frequency pumping are also used to obtain polarized targets in nuclear and particle physics, an investigative area exciting much current interest.

Moreover, the experimental and theoretical analysis techniques per-fected in nuclear physics for the study of elementary collision phenomena are now being exploited in atomic and molecular physics. Studies of resonance phenomena, originally exported to nuclear and particle physics, have been refined and extended and are once again being applied to atomic and molecular problems. It is not surprising, in view of the pervasiveness of electromagnetic interactions in determining the structure of all commonly known materials, that atomic physics also interacts strongly with technology.

The traditional physics subfields of acoustics, fluid mechanics, and optics have particularly close ties with atomic, molecular, and electron physics. Brillouin, Rayleigh, and concentration scattering have yielded new information about damping and kinetics in fluids, and the venerable science of optics has been rejuvenated by its contacts with this subfield. Holog-raphy, photon statistics, and the study of the concept of coherence have progressed rapidly in the past five years. Ultrahigh-resolution spectroscopy by means of correlations in photon arrival times has been developed. The field of nonlinear optics is approaching a peak of activity, and many new industrial applications appear likely.

This subfield also has strong ties with chemistry. Colliding-beam techniques have greatly advanced the study of low-energy atomic and molecular collisions. It is now possible to study a chemical reaction, not as a statistical thermodynamic average but with details about individual rotational and vibrational states. Experiments involving elastic, inelastic, and reactive scattering of atoms and simple molecules have led to an evaluation of interatomic forces and other phenomena, and even to the angular distribution of the products of elementary chemical reactions. The existence of relatively long-lived complexes or reaction intermediates in certain systems has been demonstrated. Evidence of the importance of the relative orientation of the colliding partners in a chemical reaction has been acquired. Obviously, atomic, molecular, and electron physics substantially overlaps chemical physics and physical chemistry.

In addition, atomic, molecular, and electron physics interacts strongly with plasma physics, astrophysics, and atmospheric physics. As men-tioned earlier, atmospheric physics is determined by collisional rate processes involving electrons, atoms, ions, molecules, and electromagnetic radiation, as is also the study of the earth ionosphere, radiation belts, solar and stellar atmospheres, supersonic flight, shock waves, space vehicle entry, and the like. To draw a line between this subfield and plasma physics or space physics is difficult. In fact, a rigid definition of boundaries is neither useful nor meaningful. The rapidly expanding amount of basic data available on highly ionized atoms and collisions between more energetic particles obviously are of interest to all the subfields mentioned above. Beam-foil spectroscopy is proving extremely useful in this context.

Atomic, molecular, and electron physics has always been important to the interpretation of astrophysical phenomena. Recently developed observing techniques are opening new regions of the electromagnetic spectrum in which the universe can be observed; greatly improved precision in the basic data of atomic, molecular, and electron physics is required for the interpretation of these observations. In addition, deeper understanding of the enormous variety of processes that can occur in the universe is required. This subfield is of obvious relevance to cosmology. To cite one example, optical observation of the relative absorption by the two lowest rotational levels of cyanogen was the first measurement of what may be the temperature of the blackbody radiation remnant of the primordial explosion that is believed to be the origin of our present universe.

Interaction with Technology and Society

Some examples of the interaction of this subfield with technology and society were mentioned in connection with present communications capability. It should be noted that entirely new technologies are emerging in, for example, thermal imaging and communications (see Figure 4.33). Society's continuing need for improved channels of communication, with ever-increasing capacity, is reflected in the growth in capacity of long-distance telecommunication links (see Figure 4.34). Improved individual communication media could alleviate travel problems in commerce, industry, government, and other enterprises and also could enhance entertainment, education, and recreation. The picturephone has come into being and probably will be widely accepted, thus necessitating a large increase in communication channel capacity, which can be provided by laser beams. Three-dimensional television, using holographic techniques, is more distant. However, holography has important applications in information storage and is used industrially for the detection of small mechanical deformations of large objects, for example, automobile tires. (Several of these applications are discussed in greater detail in the section on Optics.)

More immediate applications of light sensing and optoelectronics are in short-range control and guidance. Light radar and range finding are already

FIGURE 4.33 Crystal of lead–tin telluride. Magnification $\times 60$. Lead–tin telluride forms the basis of a new semiconductor device sensitive to infrared radiation of wavelength 8–14 μm, developed by Plessey. With a detectivity $D^* = 2 \times 10^{10}$ cm Hz$^{\frac{1}{2}}$ W^{-1} and a response speed of 10 nsec, the device, which operates at $-77°$C, opens a new field of technology in thermal imaging and communications, where it can be used to produce more noise-free systems through heterodyning laser beams. [Photograph courtesy of Stereoscan Micrograph—Cambridge Scientific Instruments Limited, England.]

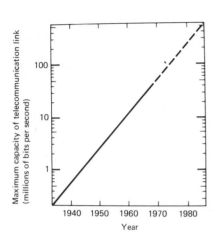
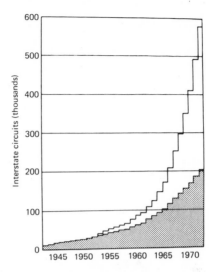

FIGURE 4.34 Increase in capacity of a single installed long-distance telecommunication link (left). Increase in number of long-line circuits in the Bell System (right). The upper curve shows thousands of interstate voice channels in the United States. The lower curve shows channels used for conventional telephone conversations. The difference between the lower and upper curves represents equivalent voice circuits for data transmission, facsimile, television, and special services. [Source: J. Martin and A. R. D. Norma, *The Computerized Society* (Prentice-Hall, New York, 1970).]

in use and may become important in air-traffic control. Public transportation, an increasingly important problem of the coming decades, also would benefit from better monitoring and control devices.

Although the conservation of resources and protection of the environment are primarily matters of economics and priorities in public policy, atomic, molecular, and electron physics can contribute to their solution through better instrumentation, which is essential for the development of the effective control devices that will inevitably be needed to implement an antipollution program. Laser-beam probing of the atmosphere and of smokestack exhausts is a new method for monitoring individual chemical constituents through high-resolution absorption and Raman spectroscopy and by Rayleigh scattering from dust particles. Measurement of temperature and water-vapor profiles in the atmosphere are essential to improved weather prediction and to possible future weather-modification programs. In the longer range, conservation of resources and preservation of the environment will require extensive and imaginative use of new technology.

This subfield also contributes, through better instrumentation, to the biomedical and health fields. Electron microscopy, holography, and various spectroscopic techniques are important analytic tools that require continuing improvement (see Figures 4.35–4.37). Fiber-optics cardioscopes, laser retina welding and cauterization, measurement of blood flow, automated spectroscopic analysis of body fluids—these all offer further possibilities for beneficial application.

It should not be forgotten that one of the most important diagnostic and clinical tools at the disposal of the medical profession is the x ray. Recent development of image intensifiers and computer techniques for image enhancement have vastly increased the potential for the use of x rays without dangerous radiological side effects. Imaginative and innovative use of x-ray diffraction also has contributed to the study and description of the structure of large biological molecules, as recounted in, for example, J.

FIGURE 4.35   A three-dimensional view of a cancer cell magnified 3000 times shows how these cells reach out to engulf neighboring cells. [Photograph courtesy of E. J. Ambrose, Chester Beatty Research Institute, London, England, and the Cambridge Instrument Company, which provided the stereoscan microscope.]

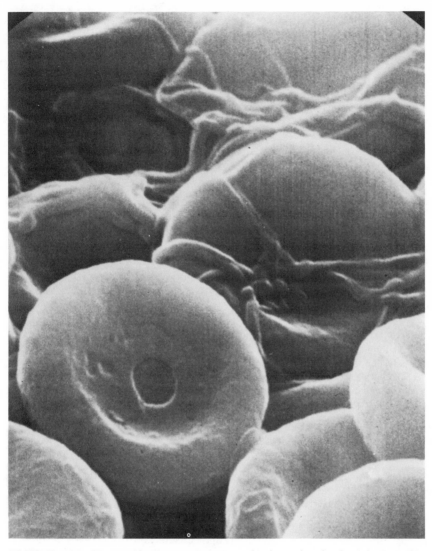

FIGURE 4.36 Human blood was allowed to clot in moist air, fixed in formalde-hyde, and viewed in the scanning electron microscope. The sample is magnified 15,000 diameters. The disks are red blood cells, held in a meshwork of fibrin strands. The cells are somewhat shrunken. [Photograph by L. McDonald provided through the courtesy of T. L. Hayes, University of California at Berkeley.]

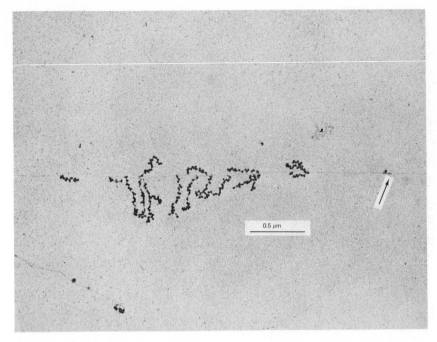

FIGURE 4.37   Visualization of bacterial genes in action by electron spectroscopy. Genetically active and inactive portions of *Escherichia coli* chromosomes are shown. The polyribosomes attached to the active segments exhibit imperfect gradients of increasing length. The shorter, most distal polyribosomes may have resulted from m-RNA degradation. The arrow indicates putative RNA polymerase molecules presumably on or very near the initiation site of these active loci. [Photograph courtesy of O. L. Miller, Jr., B. A. Hamkalo, and C. R. Thomas, Jr., Oak Ridge National Laboratory.]

Watson's *The Double Helix*.   More generally, x-ray diffraction techniques continue to play a central role in crystallography and in much of the research on the physics of condensed matter.

Atomic, molecular, and electron physics also makes major contributions to national defense and space programs.   Lasers and masers, for example, are used in radar defense, nighttime visual surveillance, bomb sighting, and satellite tracking (Figure 4.38).   Molecular physics and infrared spectroscopy are employed in the detection and tracking of rockets and in studying space-vehicle and re-entry problems.   The use of high-power laser beams for energy transmission in space, for antiballistic missile (ABM) defense purposes, and for possible ignition of thermonuclear plasmas also has received much attention.   Realizing the relevance of atomic, molecular, and

FIGURE 4.38 This huge horn antenna, weighing 380 tons, is used for satellite communication experiments at Andover, Maine. Small dish antenna in upper right of photograph was recently installed for special tests with the Communication Satellite Corporation experiments. [Photograph courtesy of American Telephone and Telegraph Corporation.]

electron physics research to questions of national security, the Department of Defense and the Atomic Energy Commission have supported a large fraction of the work in this subfield.

Future Activity

In the coming decade, much activity in a previously rather inaccessible region of the electromagnetic spectrum, the far infrared or submillimeter wavelength, will take place. Laser techniques have sparked a revolution in infrared spectroscopy that will have an impact on infrared molecular astronomy, thus continuing a historic trend as optical and microwave techniques finally converge and overlap. Extension of the new techniques of laser and quantum electronics deeper into the ultraviolet region will receive increasing emphasis in coming years. Chemical and biological

[115]

laboratories are adopting physics-based methods and applying them in new ways. Nonlinear spectroscopy, made possible by these techniques, has just begun and promises exciting results. Its present state could be compared with that of radiowave and microwave spectroscopy in 1950. The huge development of these fields between 1950 and 1960 is well documented, and a similar development can be expected for nonlinear and quantum optics during the 1970's.

Work in beam-foil spectroscopy is increasingly important for the interpretation of observations of plasmas and astrophysical phenomena.

Continued progress in the making of precise measurements and the development of more accurate standards of length and time can be expected. These, in turn, will have applications in navigational systems and precision micromachining operations and will increase the sensitivity of seismic and other such monitoring devices. Direct time measurement recently has been extended to the region between 10^{-13} and 10^{-11} sec. Consequently, details of intermediate products in chemical and biochemical reactions may be studied directly; this capability can lead to extremely valuable information about the mechanisms involved.

The colliding atomic and molecular beam techniques have attained a level of refinement that permits the details of chemical reactions and atmospheric and astrophysical collision processes to be studied in the laboratory. An entire new line of research is now possible and is undergoing rapid growth throughout the United States.

These various opportunities are also widely recognized outside the United States, notably in the United Kingdom, West Germany, France, and the Soviet Union. These and other European nations have devoted a somewhat larger fraction of their total physics effort to atomic, molecular, and electron physics than has the United States. Their research output in this subfield relative to that of the United States reflects this emphasis. Whereas in many other subfields of physics, notably high-energy and nuclear physics, the United States holds a clear lead, it does not hold such a lead in this subfield. (Research effort in this subfield in the United States is about on a par at the present time with that in other countries.)

Distribution of Activity

The central position of this subfield in science and technology and its many direct interactions with other subfields of physics, other sciences, and technology are apparent in the distribution of its workers. Atomic, molecular, and electron physics accounts for approximately 6.5 percent of the physics PhD population, as reported in the 1970 National Register of Scientific and Technical Personnel. Approximately one half of these 1,964 physicists work at educational institutions, one fourth in industry, and one fifth in government laboratories and research centers (see Figure 4.39). A special characteristic of this subfield, in contrast to nuclear and elementary-particle physics, is the absence of large permanent installations; consequently, the cost of research per unit of scientific output is relatively low. Atomic, molecular, and electron physicists combine a variety of skills and, in contrast to some of the other subfields of physics in which greater specialization and division of labor are necessary, are often well versed in both theory and experiment. Not only is this subfield well suited to the training of future scientists, but it also acquaints them with problems and techniques relevant to the demands of a technological society.

In 1970, approximately 3 percent of federal funds and 7 percent of industrial funds used for basic research in physics were allocated to this subfield. A substantial amount of research in atomic, molecular, and electron physics is supported by sources outside the subfield. If included, these might double the above percentages (Figure 4.40).

Problems in the Subfield

Because of its relevance to defense questions and to technology, atomic, molecular, and electron physics has received a large share of its funding

ATOMIC, MOLECULAR & ELECTRON PHYSICS

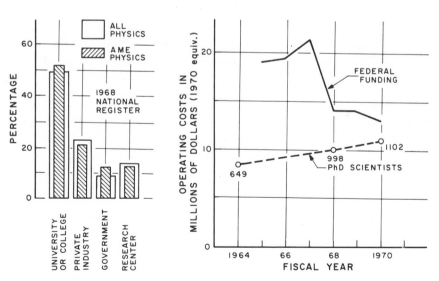

FIGURE 4.39 Manpower, funding, and employment data on atomic, molecular, and electron physics, 1964–1970.

[117]

ATOMIC, MOLECULAR & ELECTRON PHYSICS

PHYSICS MANPOWER IN 1970

N = 36,336

FUNDS FOR BASIC
RESEARCH IN PHYSICS IN 1970

FEDERAL
INDUSTRIAL

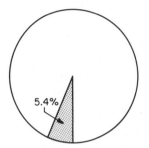

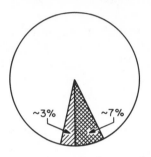

FIGURE 4.40   Manpower and funding in atomic, molecular, and electron physics in 1970.

from the Department of Defense (DOD) and from industrial sources. Even though support of relevant research should be less susceptible to cuts, a general and rather rapid decline in DOD and private industrial funding of all types of physics research appears to be under way. This trend could greatly affect the conduct of atomic, molecular, and electron physics in university settings. Support from the National Science Foundation (NSF) in this subfield, especially at the larger academic institutions, has been low. Although some correction of the distribution of NSF funding among physics subfields has occurred in recent years, it is not sufficient to offset the drastic reductions now taking place in the support of this subfield by other agencies. Federal agencies concerned with pollution, health, and transportation, which would reap long-range benefits from a vigorous program in atomic, molecular, and electron physics, should assume greater responsibility for its support.

There has been an increasing need for sophisticated data-processing equipment. Molecular-beam apparatus now requires a mass spectrograph as a standard detection instrument, and optical spectrographs, costing $2000, have been replaced by integrated spectrometers that cost ten times as much. The need for updating equipment is felt most severely at the universities. Unless a substantial increase in support for research in this subfield at educational institutions is forthcoming, most of the equipment in these institutions will become obsolete within five years. If support were to remain at its present level, a choice between continuing most small-scale efforts or sacrificing many of them to allow the conduct of a few more-expensive major experiments would be necessary. No single research area would deteriorate entirely, but general progress would be sluggish. Equipment could not be upgraded, and a general erosion of experimental facilities would begin.

> Again matter is a relative term; to each form
> there corresponds a special matter.
>
> ARISTOTLE (384–322 B.C.)
> *Physics*, Book II, 194

PHYSICS OF CONDENSED MATTER

Introduction

Condensed matter consists of solid as well as liquid, glassy, and other amorphous substances. All these substances have in common atoms so closely packed that interactions among atoms play an important role in determining their properties. The largest class of such substances is that of the crystalline solids. Their mechanical, electrical, magnetic, optical, and thermal properties have been investigated and now are understood in a general way. Such properties have been of interest since man first concerned himself with the advantages of bronze over stone or the puzzle of why bits of iron cling to a lodestone. Condensed matter was the focus of physics and much of chemistry in the nineteenth century.

Historical Background

Modern physics of condensed matter has several roots: the discovery of the structure of the atom in the early part of this century; the discovery at about the same time of quantum mechanics, which made possible the quantitative understanding of the behavior of atoms, and their constituents, the nucleus and its electrons; and the discovery in 1912 of x-ray diffraction, which gave the first quantitative information on the ordered arrangements in other forms of condensed matter. In the late 1920's and during the 1930's, the new tool of quantum mechanics was applied with vigor to develop a more complete picture of condensed matter, and quantitative or semiquantitative understanding of the more prominent physical properties emerged rapidly. Today solids are classified according to the dominant binding forces that hold them together (molecular, metallic, ionic, or

[119]

covalent bonding) and according to their electrical properties (conducting, semiconducting, or insulating). Some substances move from one part to another of the latter classification, with changes of temperature, pressure, or magnetic field, or with variations of purity. Indeed, the range spanned by the physical properties of condensed matter is probably the largest encountered in physics; as an example, the resistivity of common substances ranging from insulators to metals varies by more than 10^{10}.

Extremes of temperature and extremes of pressure have revealed astonishing variants on these usual states of matter. Thus, at very low temperatures some solids abruptly lose all resistance to the passage of electricity. This characteristic is called superconductivity. Although it was first discovered in 1911, at which time it was thought to be a property of only a small number of simple metals, it has been shown in the last decade that a remarkable variety of alloys and intermetallic compounds, and apparently some semiconductors as well, exhibit the property. A theoretical explanation of the phenomenon was not found until 1957, and many aspects of superconductivity are not yet properly understood.

A somewhat analogous loss of all resistance to flow (and other anomalous attributes) occurs in the liquid form of ordinary helium when this substance is cooled to within $2°$ of absolute zero. This condition is called superfluidity. Superconductivity and superfluidity depend on the quantum mechanical, as distinct from the Newtonian or classical, behavior of aggregates of atoms and electrons.

Mechanical properties of solids, such as strength, hardness, elasticity, plasticity, and the like, depend on cohesive qualities related to interatomic binding forces and also on certain characteristic imperfections in the ideal lattice structure of the solid. These imperfections, or lattice defects, have been the subject of intensive study in the last 25 years and now are generally understood; however, many parts of the picture, including most of its accurate quantitative features, still are lacking.

Similarly, electrical, magnetic, optical, and thermal properties of solids also are understood in general terms, and much quantitative knowledge has been achieved. The richness of the phenomena that matter can present, however, is enormous, and it is possible that today's understanding will one day be considered as crude and naive as that of half a century ago now appears.

Recent Developments

Many important developments have occurred in the study of condensed matter in recent years. Among them are advances in the understanding of the electronic structure and the elementary excitations of solids; the vast increase in knowledge of macroscopic quantum systems (super-conductors and superfluids) and their interplay with other systems; the improvement in the ability to produce many substances in states of great purity and crystalline perfection; experimentation with matter under extreme conditions and the resulting increased theoretical understanding of phenomena observed under such conditions; the achievement of broad understanding of lattice defects; the beginning of a quantitative science of amorphous materials; and the much deeper understanding of cooperative and many-body effects in condensed matter. In addition, there have been many applications of these and earlier developments to other sciences and to technology.

The electronic structures of a large number of simple crystalline solids have now been calculated from first principles in considerable detail. These calculations provide a good account of the binding and elastic properties of the substances. The motions of the carriers of electric charge have been studied, and transport properties are becoming rather well understood. Theoretical information has been checked with a variety of sophisticated experimental techniques—cyclotron resonance, de Haas-van Alphen experiments, magnetoacoustic experiments, optical absorption measurements, photoemission measurements, and many others. From this work a remarkably broad and detailed knowledge of electronic properties of a variety of pure substances resulted. This knowledge played a vital role in the design of solid-state electronic devices, solid-state lasers, and many other technological developments. This information also gives the necessary background for understanding solid surfaces (important for controlling corrosion, catalysis, adhesion, and other surface phenomena) and the effects of impurities, dopants, and alloying agents.

A solid can be regarded as possessing a state of lowest energy and many states of higher energy that correspond to various excitations. These are organized into elementary excitations of several kinds: the various modes of lattice vibration (phonons); excitation of individual electrons; excitations of the magnetic system, if any (magnons); collective excitations of conduction electrons (plasmons); and various combinations of these, which also can act as elementary excitations (excitons, polarons, and the like). Rapid progress in both theory and experiment has occurred in recent years in regard to the understanding of excited states of solids in these terms.

Although superconductivity and superfluidity have been known for decades, they continue to be subjects of great excitement, with important advances, both scientific and technological, taking place. The modern theoretical understanding of superconductivity dates from 1957, when Bardeen, Cooper, and Schrieffer proposed their now-famous theory. An outgrowth of this understanding was the proposal by Josephson in 1962 that a junction between two superconductors separated by a very thin insulating layer should possess anomalous properties. Josephson predicted that a current of limited magnitude could pass through such a junction with zero voltage between the two superconductors and that the magnitude of this current would be highly sensitive to magnetic fields, going periodically to zero as the field is changed. He also predicted that if a constant voltage bias were imposed across the insulating layer, an alternating current would flow with a frequency proportional to the voltage bias. These predictions were quickly confirmed by experiments and have led to the production of numerous practical laboratory instruments of unparalleled sensitivity and to better methods of measuring certain fundamental constants of nature. Superfluids also have exhibited a rich and astonishing range of phenomena in recent years, and a close parallel between properties of superconductors and superfluids has been found. The ideas involved have had important applications to theories of the atomic nucleus and the interior of certain kinds of stars.

The transistor and a number of other solid-state electronic devices require materials of unprecedented purity and perfection, for the electrical properties of semiconductors are influenced dramatically by impurities and crystalline imperfections. Techniques of producing many materials of exquisite purity, and in the form of nearly perfect single crystals, have been developed by necessity as a part of semiconductor technology. These methods have been useful in producing much better materials for research than were previously available. As a result, elucidation of the intrinsic properties of these substances and the extrinsic properties associated with impurities and imperfections became possible. These sensitive and far-reaching distinctions are a hallmark of modern experimentation in the physics of condensed matter.

An offshoot of the development of the high-purity germanium and silicon required for the device market has had revolutionary impact on nuclear and atomic physics. With adequate purity it was possible to construct semiconductor radiation detectors—in essence, solid ionization chambers—that provided energy resolutions typically one to two orders of magnitude better than those previously attainable; moreover, this paramount advantage is coupled to a wide variety of other inherent advantages including small size, elmination of high-voltage requirements, and insensitivity to magnetic fields and, when so designed, to ambient radiation backgrounds as well.

Continuing feedback into technology also occurred. Turbine blades for jet engines, for example, are now made in single crystals of metal, using well-developed techniques for crystal growth. These single-crystal blades have superior ductility and strength, because the boundaries between crystallites, present in ordinary metals, give rise to brittleness.

Interesting modifications occur in many substances when they are subjected to very high pressures. The material in the interior of the earth is subject to such pressures, which can now be attained in laboratory experiments. Static methods are used to reach pressures of several hundred thousand atmospheres. Shock waves, produced by chemical high explosives or other means, produce transient pressures into the range of millions of atmospheres. Theories of interatomic binding are tested by such experiments. Extension of the theoretical treatment to still higher pressures gives an account of matter in the interior of stars and is necessary in completing the understanding of astrophysics. Modern ideas of the mechanisms of crystal growth have guided the production of useful substances by means of high pressures combined with high temperatures. Thus diamonds have been made artificially, first of industrial quality, more recently of gem quality.

Very high magnetic fields have been developed for laboratory purposes. Static fields are produced in conventional coils, supplied with heavy currents, and in the coils made of recently discovered hard superconductors. Short-term transient fields of still higher strength are produced in various kinds of pulsed coils. These high magnetic fields subject the electrons in a solid to a perturbation that tests and elucidates the electronic structure of the solid. Such work has produced a vast amount of new information on electronic structures of solids.

Very high and very low temperatures have also been used to extend the knowledge of condensed matter. Ultralow temperatures are being achieved by sophisticated combinations of new methods, including the ^{3}He–^{4}He dilution refrigerator, and new kinds of cooperative phenomena are being sought. The quantization of magnetic flux in superconductors and of flow velocity in superfluids are two of the striking experimental findings relating to macroscopic quantum phenomena that have emerged recently from this work. Polarization, that is, spin alignment, of atomic nuclei has become a commonplace achievement. This is particularly interesting as a tool for the nuclear physicist. A number of new superconductors and new magnetic transitions have been found.

Lattice defects can be divided into two classes: point defects, which are of atomic scale, and extended defects, which are of macroscopic scale in one or two directions. Vacant sites on the lattice are point defects; lattice mismatches along an extended line, called dislocations, are extended defects. In many solids diffusion occurs by the hopping of vacancies from one lattice site to another, under the influence of the thermal agitation of the lattice. Catalytic effects often depend critically on the presence of lattice defects at a crystalline surface. Irradiation of metals by nuclear radiation, as in a reactor, causes important changes in properties, because of the point lattice defects produced by the radiation. Strength and plastic properties of solids depend markedly on the presence of dislocations and on point defects that impede their motions. All of these defects have been studied for almost three decades, but in the last decade the generally qualitative understanding is becoming quantitative. The variety of possible defects in different crystalline substances is enormous, and the quantitative knowledge that is in hand is but a small fraction of that which would be of immense practical value.

Glasses, liquids, and many other condensed substances are said to be amorphous; in other words, the arrangement of their atoms is only semiregular. For many years electronic properties, such as electrical conduction, as well as lattice vibra 'ons were treated by quantitative theories only in crystalline solids. Recently, these phenomena have been studied in amorphous substances, and a theory of electronic properties and lattice dynamics in such substances is beginning to emerge. Computer studies entailing the analysis of the behavior of models of amorphous substances have been of benefit as stimulants to intuition and guides to theory. Computer studies of the atomic dynamics in model liquids have been especially useful.

The sharp transition from a solid to a liquid, which occurs at the melting temperature, is an example of what is called a cooperative effect. Another example is the abrupt loss of long-range order in the magnetization of the atoms of an iron lattice when a certain characteristic temperature is reached. These phenomena depend on the cooperative interaction of many elements of a dynamical system and have been objects of study in physics for years. Recently, advances in theory have shown that many disparate cooperative phenomena are mathematically related. Many new examples of cooperative transition have been found and studied by dynamical as well as static methods. The scattering of laser light and thermal neutrons by such systems afforded new experimental information on the dynamical and static properties of these systems. Cooperative phenomena occur in what may be called many-body systems. Every system studied in the physics of condensed matter is actually a many-body system. At an earlier stage, physics usually progressed by reducing the many-body systems, conceptually, to an approximate replica in which one body at a time could be considered as moving. Although it is not yet possible to give a rigorous dynamical treatment of general many-body systems, much improvement has been made in methods of calculation so that good corrections for many-body effects can be made in an increasing array of problems. The understanding of superconductivity requires explicit allowance for such effects, and a great deal of the theoretical effort in the physics of condensed matter in the last decade has been devoted to understanding many-body corrections to earlier one-body models.

Other branches of physics have interacted recently in vital symbiosis with the physics of condensed matter. The Mössbauer effect is the phenomenon of emission of a nuclear gamma ray by an atom imbedded in a solid, without the recoil of that atom, that would ordinarily occur and somewhat reduce the energy of the gamma ray. This phenomenon combines the realms of nuclear physics (the emission of the gamma ray and its dependence on the state of the nucleus) with the freedom from recoil that is a consequence of the binding of the radiating atom into a crystalline lattice. The effect has been an important tool for learning about crystalline binding and the magnetic field (and the gradient of electric field) at the nucleus of the emitting atom. Some of this information complements that which can be obtained by the older phenomenon of nuclear magnetic resonance. Another measure of the magnetic field at the nucleus of an atom in a lattice results from observations of the angular correlation between two successive cascade gamma rays emitted by the nucleus. The phenomenon of radiation introduces lattice defects into a solid and constitutes a tool for studying lattice defects.

Still another example is the process of channeling of energetic charged particles by a crystal lattice (see Figure 4.41). In channeling, a nuclear particle, such as a proton or an alpha particle of a few million electron volts energy, when falling upon a single crystal at a direction nearly parallel to a prominent axis of the crystal, tends to be guided into a channel accurately parallel to this axis and to lose energy more slowly under these conditions than when going through the crystal in a random direction. Not only has this process added to the understanding of the process of energy loss by fast charged particles in solids (which is an old subject), it has also provided a new tool for investigating the perfection of crystals and learning the configuration of point defects in crystals. Even more recently, it has been used to measure nuclear lifetimes in the range of 10^{-15} sec, which previously have been completely inaccessible, through effective observation of flight times between lattice sites.

Relationship with Other Sciences

The physics of condensed matter has strong interactions with chemistry and metallurgy and with the broader field of materials science. These interactions are too numerous, and most are too obvious, to catalog here. Usually the solid-state physicist and the solid-state chemist are distinguishable only by their backgrounds or, occasionally, the emphasis in their work. The chemist is more often concerned with complex substances (although this difference is decreasing); methods of preparation, synthesis, crystal growth, and the like; and use of chemical methods of

channelling

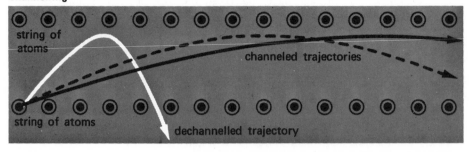

blocking

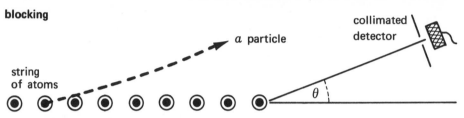

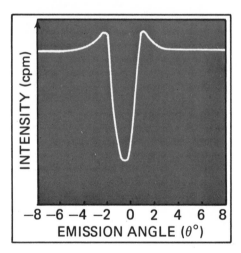

FIGURE 4.41a   Channeling of charged particles through a crystal matrix can occur only if the incident angle is less than a certain critical value. Then the incident particles suffer a series of gentle repulsions from the string atoms and remain in the channel. When the angle is too large, the particle is strongly deflected and can no longer be channeled. The opposite process is known as blocking; particles originating in the center of the string cannot be channeled. Here alpha particles are shown being emitted from radioactive radon atoms situated on lattice sites in tungsten. Along the direction of the string, which corresponds to an emission angle of 0°, the intensity of the emission goes through a minimum value caused by the blocking phenomenon.

analysis. Metallurgy has been strongly influenced by concepts of electronic structure and binding originally developed by physicists. The large and productive field of defects in metals has become increasingly the province of the metallurgist.

A few selected examples will illustrate the widespread impact of the physics of condensed matter on other subfields of physics and other sciences. The Josephson effect, as noted above, consists of an oscillating

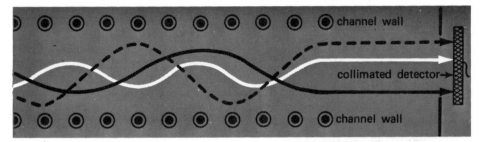

FIGURE 4.41b Position of the collimated detector imposes a simple geometrical boundary condition on the number of particles that are detected after they leave the channeling planes. Only those particles that leave the crystal in a direction very near to the plane can be detected if the detector is placed far enough away. This makes it easier to study the processes by which particles lose energy, by limiting quite drastically the number of oscillations that are observed. In the diagram above, the boundary condition allows only three oscillations to reach the detector, producing the three peaks in the energy spectrum (left). Studies of this kind, although difficult, make it possible to obtain information from the energy spectrum about the electron distribution and interatomic potentials in the individual atoms in the crystal lattice.

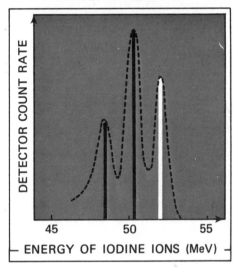

current between two superconductors in weak contact maintained at some potential difference V. The frequency of the oscillation is given by $v = 2eV/h$, where e is the charge on the electron and h is Planck's constant. From a measurement of v and of V, this relationship determines the important natural constant e/h. This method of obtaining e/h is several orders of magnitude more precise than any alternative experimental method. In conjunction with other measurements, this method has resolved a long-standing discrepancy in the quantum electrodynamic theory of the hydrogen atom and certain other elementary physical quantities. This finding reinforces belief in the validity of quantum electrodynamics, which is one of the most basic and far-reaching components of the current conception of the physical world. Thus, an understanding of the superconducting state in solids has had implications for the seemingly unrelated fields of quantum electrodynamics and atomic physics. The Josephson effect also promises to provide the basis for a very compact, high-efficiency computer memory system; such systems are being developed at the present time.

The impact of condensed-matter physics on nuclear physics is multi-dimensional. Nuclear magnetic resonance of atoms in solids, the Möss-bauer effect, angular correlation of successive gamma rays, and blocking effects in crystals are examples of effects that, through sophisticated knowl-edge of solids, allow measurement of major parameters of nuclei. Perhaps the most dramatic impact of condensed-matter physics and associated tech-nology on experimental nuclear physics, however, has been through semi-conductor detectors of nuclear radiation.

In theoretical nuclear physics, major progress resulted from the applica-tion of the Bardeen-Cooper-Schrieffer theoretical breakthrough concerning superconductivity in condensed matter. This recognition of the joint dependence of these subfields on many-body phenomena greatly enriched both. The Josephson effect between superconducting solids, discussed above, is now being actively sought in the nuclear realm in the interface between colliding superconducting nuclei such as selenium and tin.

Physicists and chemists concerned with solids have devoted much effort to the study of surfaces. In spite of this work, heterogeneous catalysis remains a highly empirical field. Condensed-matter physics has contributed techniques that promise increased understanding. Thus it is now possible to cleave single crystals in very high vacuum to produce clean and regular surfaces that can be microscopically characterized. Then, by means of low-energy electron diffraction, two-dimensional order in a monatomic layer of adsorbed gas can be studied. Recently, a new technique, ion-neutraliza-tion spectroscopy, has been developed that is sensitive to processes going on within an atomic diameter of the metal surface by ions of noble gases, and the energy of the ejected electrons is measured. From these data infor-mation about the chemical bonds in molecules adsorbed on the surface can be deduced. Such molecules differ in interesting ways from free molecules, because they are constrained by the solid and can have structures that do not occur among free molecules but are likely to be of prime importance in both corrosion and catalytic reactions.

There are many ways in which condensed-matter physics is impinging on earth and planetary physics (an example is given in Figure 4.42). Our knowledge of the earth and its history is based to a large extent on solid-state physics and chemistry. The model for the composition of the mantle is based on the phase equilibria determined for different germanate systems that can be extrapolated to the silicates, which comprise the earth's mantle. The knowledge of the response of a solid to sound waves and studies of the equation of state of liquids allow one to hypothesize a liquid core; set

100μm

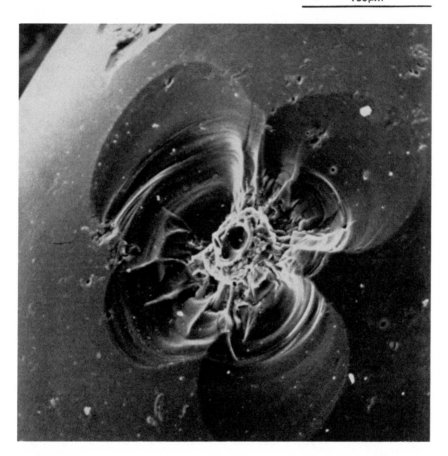

FIGURE 4.42 Apollo 11 moon rock sample. Micrometeorite impact crater on the surface of a glassy sphere showing a central crater (produced by melting) standing on a boss with irregular vertical fractures, surrounded by petal-like conchoidal (brittle) fractures. The small black areas are bubbles in the original glass which have been cut by the fracture or burst on the natural surface of the sphere. Magnification ×520. [Photograph courtesy of Stereoscan Micrograph—Cambridge Scientific Instruments Limited, England.]

a limit of the temperature and pressure in the mantle; and set some limits to the accuracy of the extrapolations that are made concerning the properties of materials in the earth.

[129]

Impact on Technology

Over the past several decades, the physics of condensed matter has had great impact on technology; its influence shows no sign of abating. Indeed, technological benefits from this subfield probably will emerge even more rapidly in the future. (Section 2 of the report of the Panel on Condensed Matter, in Volume II, presents an outline of recent developments in condensed-matter physics that have direct technological importance; Section 3 gives case histories of selected innovations; and Section 5 discusses the economic and social consequences of these technologies.)

Perhaps the single development with the most far-reaching consequences is the transistor, development of which began in 1947 with the discovery of the transistor effect by Shockley, Bardeen, and Brattain (see Plate 4. VI). An enormous number of solid-state electronic devices have evolved since. Today's large-scale integrated circuitry is even farther ahead of the original transistor than that device was ahead of the old vacuum tube (see Figure 4.43 and Plate 4. VII). Control circuits and computers of high speed, high reliability, low cost, and small size are only a few of the benefits

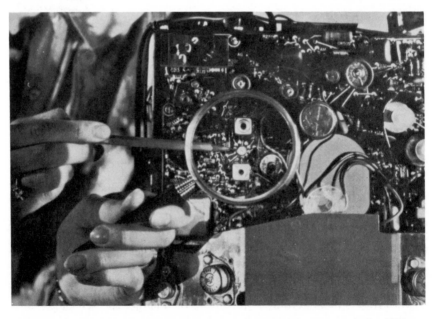

FIGURE 4.43 First commercial use of integrated circuits was made in 1966 RCA television receivers. [Source: *Science Year. The World Book Science Annual.* Copyright © 1966, Field Enterprises Educational Corporation.]

PLATE 4.VI This is the first transistor ever assembled; the year was 1947. It was called a point contact transistor because amplification or transistor action occurred when two pointed metal contacts were presented onto the surface of the semiconductor material. The contacts, which are supported by a wedge-shaped piece of insulating material, are placed extremely close together so that they are separated by only a few thousandths of an inch. The contacts are made of gold, and the semiconductor is germanium. The semiconductor rests on a metal base. Public announcement of the transistor was made July 1, 1948, by Bell Telephone Laboratories. In 1956, John Bardeen, Walter Brattain, and William Shockley shared the Nobel Prize, the highest honor in science, for their discovery of the transistor effect. [Source: Bell Telephone Laboratories.]

[131]

PLATE 4.VII Close-up of a single microminiature beam-lead circuit. This circuit was developed by Bell Labs for possible use in electronic switching system equipment. A dual five-input NAND gate, it has 10 transistors, 18 diodes, and 12 resistors. Actual size of the circuit is 0.053 in. from beam tip to beam tip. [Source: This photograph was provided through the courtesy of Bell Telephone Laboratories.]

that these devices have made possible. Small chips of silicon are now made in routine fashion containing hundreds of thousands of electronic components. In the past five years, integrated circuits have come to account for sales of $500 million per year, which is 30 percent of the total semiconductor market. Recent innovations are solid-state microwave diodes and optoelectronic devices. The former, including Gunn-effect oscillators, probably will be widely used in communication and radar equipment. Optoelectronic devices include solar cells, which provide power for satellites, and light-emitting diodes, which make possible new indicator devices and display screens.

On the horizon are optoelectronic devices that include laser light sources, sophisticated large-scale computational elements, and very-large-scale memories, all fabricated within the same crystal wafer. Even more exciting is the possibility of overlapping multiple systems through the same optical channels, each operating on its own wavelength of radiation. Such devices, with ultraminiature size, microscopic power requirements, effectively infinite lifetime and reliability, and low intrinsic cost cannot fail to have at least as revolutionary an effect on society as did the introduction of computer techniques. (See Plate 4.VIII.*)

Optoelectronic techniques also hold great promise for the development of whole new generations of visual displays in the all-important man–computer interface; these can be cheaper, safer, more compact, and more convenient by orders of magnitude than the typical cathode-ray tube band systems now in use.

Magnetic materials are widely used now and comprise many substances besides the steels of transformer cores and electric motors. Ferrites, which are in effect ferromagnetic insulators, are used in television transformers and computer memories. Other new magnetic substances appear in ferromagnetic memory devices, such as tapes, drums, and disks. Magnetic materials that can be stimulated by optical means, such as laser pulses, may also have an impact on display and information-storage devices.

Magnetic bubble technology has provided a radical new way to store information in magnetic bubbles in thin, transparent, magnetic crystals and to carry out logical operations by moving the bubbles over the crystal (see Figure 4.44). Magnetic bubble memories may replace both the core and disk file memories in computers and electronic central offices and interface directly with fast semiconductor devices. They may prove to be a very fast, compact, and inexpensive way to store and process data.

Superconducting materials have come into use for a variety of specialized needs, particularly for providing high magnetic fields for laboratory purposes. The greatest impact of these new materials, and of others under development, probably lies ahead, with power transmission lines (see Plate 4.IX), transformers, motors, mass-transport levitation and guide-

* Plate 4.VIII is not reproduced here;
the reader is referred to
Physics in Perspective, Volume I.

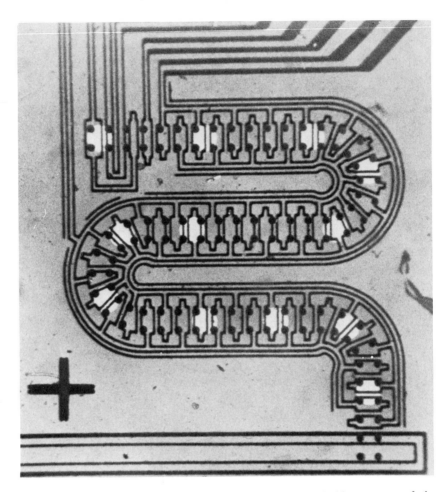

FIGURE 4.44   Tiny computers and electronic telephone switching systems of the future may accomplish counting, switching, memory, and logic functions all within one solid magnetic material, employing new technology now in exploratory development at Bell Telephone Laboratories. Looking more like a block diagram or a flow chart, this actual circuit, a photolithographic pattern on the surface of a sheet of thulium orthoferrite, can move magnetic bubbles (large white dots) through a shift register. The magnetic bubbles are 0.004 in. in diameter. [Source: This photograph was provided through the courtesy of Bell Telephone Laboratories.]

PLATE 4.IX   A 40-ft section of high-voltage (450,000 V) cryogenic transmission cable undergoing testing at the General Electric Company.

[135]

ways, and computer memories being potential areas for future application. Ultrasensitive electric and magnetic measuring devices, another application of superconductors, already have been developed.

Of crucial importance to any future implementation of superconducting technology, as, for example, in transmission of electrical power or in new computer memory configurations, is the maintenance of an adequate supply of helium. Helium is unique in its combination of unusual properties and critical uses. It is essential for cryogenics, superconductivity, some types of breeder reactors, and the space program. Moreover, large components of contemporary physics research—in condensed matter, elementary particles, and nuclear physics—are entirely dependent on liquid helium to achieve adequately low working temperatures. According to available estimates, this irreplaceable resource is in short supply, yet it is being wasted in alarming fashion.

Solid-state lasers of various kinds have had tremendous impact in many fields and promise further developments, including light-beam communication systems with a carrying capacity far greater than any system available today.

Secondary emission is a phenomenon from condensed-matter physics that has long found application in the electron multiplier. Image-intensifying devices based on this effect provide the starlightscope, a medical x-ray technique that permits much lower exposure of the patient to radiation and sensitive detectors that improve the seeing power of astronomical telescopes by large factors. New compounds have made possible sensitive detectors for the infrared. One example of their use is in the military snooperscope.

Fundamental discoveries in polymers, crystallization, morphology, radiation damage, point defects, dislocation, diffusion, and annealing have aided the development of a host of new materials: new rubbers and new composites, alloys for use in nuclear reactors, high-temperature and high-strength material, and new steels and alloys. Better materials are being developed for biological and medical uses. These include membranes for artificial organs and strong inert materials for replacement of joints and for heart-valve implants. Although progress has been made, the potential uses of biocompatible materials are just beginning to be realized.

Future Activity

Forecasting the trends of any active scientific field is difficult and uncertain. However, based on present activity, it appears reasonable to predict that the areas of vigorous activity and rapid progress in the physics of condensed matter in the next several years will include the following:

[137]

1. Surfaces and interfaces. New experimental techniques and increased theoretical attention should result in new breakthroughs.

2. Optical properties of solids. Lasers and new radiation sources, including synchrotrons for the far ultraviolet, have reinvigorated this field. Nonlinear optical properties of solids, exciton absorption and luminescence, and optical properties in the far ultraviolet and soft x-ray region offer a wealth of new opportunities.

3. Scattering studies on solids and liquids. New techniques and more intense sources of neutrons and photons have opened entirely new areas of research. This work has large potential for providing fundamental information on liquid and solid structure, and through that the potential for the development of both improved and entirely new devices.

4. Complex crystalline substances. Most of the substances that have been extensively investigated are relatively simple ones in which phenomena are not hopelessly complex and well-characterized specimens have been available. Silicon, germanium, the simple metals, and alkali halides have been among the most commonly studied solids. Not all useful solids are simple, however; the number of little-studied substances vastly exceeds the number of well-studied ones. Attention is turning now to a much wider and more complicated realm of substances, with three, four, or many atoms per unit cell.

5. Disordered condensed materials. The conceptual framework of the electron theory of solids and of lattice dynamics is being extended to include disordered materials. New materials in this category are being discovered and synthesized with new techniques. The development of useful electronic devices from such substances is likely.

6. Electrons and phonons as elementary excitations in solids. The motion of free charge carriers in solids and its elaborations offer rich ground for discovery and innovation. Bulk negative conductance in gallium arsenide, many-body effects in solid-state plasmas, propagation of helicon waves in high magnetic fields, and plasma instabilities arising from nonequilibrium conditions are examples. Investigation of the vibrations of crystal lattices (lattice dynamics) also continues to be an active area. Lattice vibrations play an important role in a wide variety of phenomena, for example, superconductivity, ferroelectricity, and antiferroelectricity. Moreover, significant advances have occurred in experimental techniques, including inelastic scattering of neutrons, Raman scattering of laser light, electron-tunneling spectroscopy, and ultrasonics.

7. Channeling, blocking, and related phenomena. These phenomena are finding important applications in studies of lattice structures and lattice defects. They provide insight into radiation damage and have important applications in nuclear physics, such as the measurement of very short lifetimes of nuclei. The related technique of ion implantation is becoming

increasingly important in the production of complex semiconductor devices.

8. Low-temperature phenomena and superconductivity. An important frontier of solid-state phenomena lies at the extreme low temperatures. New experimental techniques have pushed this frontier back, and new phase transitions and novel collective phenomena are sure to be found. Better understanding of superconductivity and superfluidity will follow.

9. Applications to astrophysics and extreme states of matter. Many of the principles in the theory of condensed matter at extremely low temperatures can be extrapolated to white dwarfs, neutron stars, and other problems in astrophysics. Along somewhat different lines, laboratory work in the extreme ultraviolet is called for in connection with the discovery of stellar x-ray sources. Matter under very high pressure, and also under very high magnetic fields, is being more widely investigated, both theoretically and experimentally.

All the foregoing areas appear ripe for vigorous exploitation, and unforeseen developments in this rich and rapidly moving subfield will undoubtedly play a prominent role in the years immediately ahead.

Distribution of Activity

The study of the physical properties of condensed matter and the search for understanding of these properties constitute, by a substantial margin, the largest subfield of physics. Condensed-matter physicists account for about 25 percent of the physics PhD population, as reported in the 1970 National Register of Scientific and Technical Personnel. Figure 4.45 shows the distribution of these physicists among employing institutions. About 40 percent of these work at educational institutions, 38 percent in industry, and about 18 percent in government laboratories and research centers.

The physics of condensed matter differs sharply from most of the other subfields because industrial support of the subfield exceeds government support. Much of the industrial effort is applied in character, but there is a continuous gradation between the application-oriented efforts and the basic efforts, and the industrial contributions to the basic side are highly significant. According to recent estimates, approximately 53 percent of the basic research in the subfield takes place in universities, 20 percent in government laboratories, and 25 percent in industry. About 75 percent of the more applied work is found in industrial laboratories. The industrial component of support cannot be assumed to act as a ballast, rising when the federal support decreases and decreasing after federal support rises. In fact, the trend over the past 20 years has been for the two to move synchronously.

The typical research project is relatively small in cost; and large devices

CONDENSED–MATTER PHYSICS

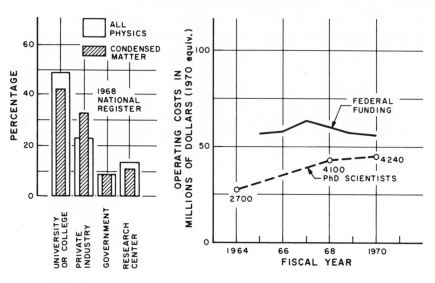

FIGURE 4.45 Manpower, funding, and employment data on the physics of condensed matter, 1964–1970.

do not consume a major portion of the funds. The contrast with elementary-particle physics and nuclear physics, in this regard, is particularly sharp. The diversity of the subfield manifests itself in the number and variety of research topics being pursued, as well as in the interplay of fundamental and applied research. The separation between basic discoveries and applications in condensed matter is less distinct than in many other areas of physics.

In 1970, approximately 12 percent of federal funds and 75 percent of industrial funds for basic research in physics were allocated to research in condensed-matter physics (Figure 4.46).

Problems in the Subfield

As noted before, the physics of condensed matter consists for the most part of many small to medium-sized research efforts widely dispersed in universities, industries, and governmental and national laboratories. Such a diverse effort, which has few individual projects with an immutable claim to survival, is at a disadvantage, in a time of retrenchment, in the compe-

CONDENSED MATTER

PHYSICS MANPOWER IN 1970

N = 36,336

FUNDS FOR BASIC
RESEARCH IN PHYSICS IN 1970

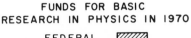
FEDERAL
INDUSTRIAL

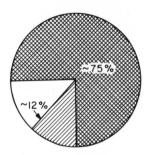

FIGURE 4.46   Manpower and funding in the physics of condensed matter in 1970.

tition with other subfields built around large projects that represent major items in the budget. In these circumstances, the small projects tend to become unduly eroded. At the same time, the condensed-matter subfield is developing an increasing need for large facilities: sources of intense magnetic fields, sources of high pressures and high temperatures, large computers, more-intense beams of thermal neutrons, improved electron microscopes, improved synchrotron sources of far-ultraviolet radiation, expanded facilities for preparation of pure materials, and special sources of particle radiation. These are not being provided in the United States at present, and, indeed, some existing facilities have recently been closed. Moreover, the increasing sophistication of the best laboratories in the subfield works to the increasing disadvantage of the smaller laboratories, which have always played a significant role.

To an increasing degree, research in condensed-matter physics, and other subfields of physics, is based on a wide array of experimental techniques that are applied to a single scientific question; the smaller laboratories cannot mount such broad attacks. Regional facilities that supplement the apparatus available to physicists in many small laboratories are needed, but new facilities of this character are not being set up. University departments with strong specialties in the subfield are now severely hampered by lack of funds for fellowships and for the research costs incidental to training.

[141]

> All men by nature desire to know. An indication
> of this is the delight we take in our senses; for
> even apart from their usefulness they are loved
> for themselves; and above all others the sense
> of sight.
>
> ARISTOTLE (384–322 B.C.)
> *Metaphysics,* Book I, 980

OPTICS

Introduction

Optics is a basic and applied subfield of physics traditionally divided into physical optics, geometrical optics, and physiological optics. Although the major emphasis has always been on visible light, optics usually is generalized to include the techniques and phenomena of electromagnetic radiation extending from the far ultraviolet to the far infrared and occasionally to x rays, microwaves, and even electron optics.

Classical nineteenth-century optics emphasized optical instruments such as the microscope and telescope; visual phenomena such as the sensation of color; and physical optics of interference, diffraction, spectroscopy, polarized light, and crystals. Modern optics has added a tremendous number of intricate new phenomena produced by the interaction of light with matter. Holography, photon counting, the biochemistry of vision, photography, photoconduction, light-emitting diodes, photoemission, and the laser are examples. Modern optical instruments include luminescent display panels, image intensifiers, electrooptic light modulators, tunable lasers, guided waves, amplifying fibers, pulse compressors, image-enhancement systems, tracking devices, and combinations of these and other new tools.

Throughout the nineteenth century, optics was a central subject of physics and commanded the attention of the greatest physicists of the time, but the skill and insight of these great physicists seemingly exhausted the topics to which they addressed themselves. For example, the Abbé theory of the microscope established a clear-cut limit to the resolving power of the instrument and, when commercial instruments reached that limit, the subject was seemingly closed. The major exception in optics—the subject that was not exhausted—was the field of atomic and molecular spectroscopy. This subject, which still challenges many physicists, now constitutes a separate body of knowledge in atomic, molecular, and electron physics, though it still is closely related to optics.

Much of the stimulus to optics in the last two decades developed as a deeper understanding of the limiting factors in many experiments and devices was achieved. Many problems throughout science were found to be "optics-limited." That is, the speed or accuracy with which a measurement could be made, a device controlled, an object detected, or a chemical analysis completed often was limited by fundamental optical problems of intensity, resolving power, stability, or photon statistics. The search for ways to overcome these limitations led to extensive programs of applied optical physics. In addition, the astonishing attributes of the laser made it productive to re-examine all these optics-limited situations. For example, it is now possible to determine with unprecedented precision the Raman spectrum of a few milligrams of a water-soluble biological compound using a laser as a light source. Without the laser, only massive samples can be analyzed. A laser beam shining on a beam of high-energy electrons suffers Compton scattering and is shifted in wavelength to the gamma-ray region without loss of polarization; the scattering cross section is low, but the intensity of the laser beam is high. This technique has already provided a powerful new tool for elementary-particle physicists.

There appears to be no shortage of frontier areas in optics at this time, partly because so many situations are still optics-limited, so that each new fundamental development has a rapid and direct effect on applied optics. The need for re-examination of long-established assumptions has had many important consequences in optics. Abbé and Rayleigh had assumed that the purpose of the microscope was to see small opaque or self-luminous objects, thus they defined the figure of merit according to this purpose. Zernike chose a different figure of merit—the ability to see small transparent living objects. He discovered the phase microscope, a simple method of great subtlety, and won a Nobel Prize.

The pattern continues today. The majority of optical scientists (and there are some 6000 members in the Optical Society of America) currently work on practical problems in optical engineering or applied optics. But questions arise frequently that challenge present understanding of the basic physics, demonstrating that there are not only unanswered questions but that the classical questions often are not the correct ones.

Spectographic instruments provide an important contemporary example. Every physics student is taught something about the wavelength resolving power of a spectrometer and shown how it depends on the properties of the dispersing element, but the arguments that he hears do not deal with the time devoted to the observation or the signal-to-noise ratio. The modern Fourier-transform spectrometer, built by Pierre and Janine Connes, has produced planetary spectra with 100 times greater resolving power and 10 times better signal-to-noise ratio than the best prism spectrometers. Much of their success arises, of course, from their technical skill, but the underlying physics is beautiful and profound. Their new atlas of planetary spectra shows more detail in the absorption spectra of the atmosphere of

Venus than is shown in corresponding spectra of the earth's atmosphere that have been obtained with prism spectrometers (so-called solar atlases).

New Developments in Optical Physics

Although optics is now predominantly an applied science, advances in basic optical physics continue to occur. By far the most important recent advances in optical physics are the laser and the new devices and techniques that it has made possible. The invention of the laser can be credited to atomic and molecular physicists, but optical specialists played a basic role from the beginning. The availability of copious coherent light has made possible a wide variety of new optical techniques such as the hologram and related image-processing methodology (see Figure 4.47), and the high intensity of lasers has led to the development of the exciting new work in nonlinear optics.

Currently there is no apparent slowing of the pace of development associated with the laser and laser devices. Laser techniques are being extended into new wavelength regions; lasers are being made tunable to different wavelengths; and new techniques are being developed for deflecting and modulating laser beams and producing and controlling extremely short pulses of light. (See Plate 4.X).

A wide variety of special situations has resulted from the use of this new tool. The ability to produce ultrashort pulses of light, for example, has opened the time domain in chemical kinetics to investigation in a way that was impossible with more conventional sources; now the sequence of events can be unraveled and followed. Very short pulses can also produce fantastically high-power densities and field strengths—so high that even nuclear forces may be affected. A completely new tool thus becomes available for studies of nuclear fusion as well as for other purposes. Eventually, these high photon densities may allow the direct experimental observation of the scattering of light by light; however, a vacuum more perfect than any now obtainable will be necessary.

The frequency stability of even a simple laser gives light of quite remarkable purity, but stabilized lasers can be built whose frequency is fixed to one part in 10^{11}. The application of this stability to phenomena now thought to be well understood will certainly result in surprises. Meanwhile, even much simpler lasers in unequal path interferometers allow measurement of strain, tilt, and shear in the earth so that earthquakes can be studied and perhaps predicted with new accuracy.

A particularly significant development arising from the laser is the invention of the modern hologram in 1962 by Leith and Upatnieks. Gabor had formulated the conceptual base for the hologram in 1949, but it was

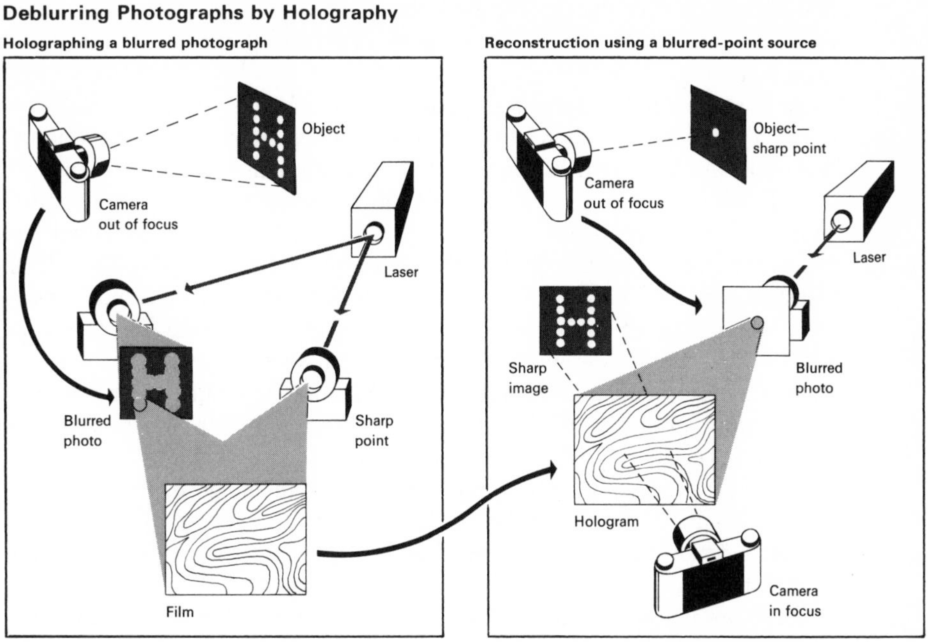

Deblurring Photographs by Holography

Holographing a blurred photograph

Reconstruction using a blurred-point source

FIGURE 4.47 A photograph taken out of focus can be sharpened with the use of holography. First, a conventional hologram is made, above left, as laser light from a sharp source simultaneously interferes with light from each blurred point in the scene. If the hologram were then illuminated with a sharp light source, the same blurred scene would be reconstructed. In a technique created by George Stroke of the State University of New York at Stony Brook, the image is instead reconstructed, above right, using a point of light made out of focus by the same camera. Each point on the reconstructed scene then appears as sharp as the sharp light source originally used to make the hologram. [Source: *Science Year. The World Book Science Annual.* Copyright © 1968, Field Enterprises Educational Corporation.]

not until coherent light was available in copious supply that large-scale, intense holographic displays became possible. The theory of the nature of optical images has been challenged by the hologram, and a wide variety of new experimental techniques have become necessary. Throughout this decade of development of holography, optics has been stimulated constantly by the prospect of useful devices, and so substantial investments of money and manpower have been made in this subfield.

The hologram can preserve and reproduce an image, and, within certain limits, a three-dimensional image of an extended object is possible. The image need not arise from a real object; a hologram can be produced from a computer printout. Thus, it is possible to generate an image of an object

[145]

that did not exist or to tabulate the sound field around an object illuminated with coherent ultrasound or to recreate an image of the object for visual examination.

In principle, holography also is useful as a means of storing data for use in a computer or other information-retrieval system. Because the information in the hologram is not localized, scratches, dirt, or other blemishes in it do not introduce errors or destroy information. In the same way that communications engineers have learned to analyze signals in either the frequency domain or the time domain, the hologram makes it possible to transform images between object and aperture spaces. It is not always clear which type of space has practical advantages over the other, but the analysis of the hologram has increased understanding of information storage systems and offered new possibilities for extending their scope.

The scope of image storage can be increased in another respect as photochromic materials with improved properties become available. With such materials, it is possible to generate and record an image using one wavelength of light, to read it or extract information from it by using a second wavelength, and, finally, to erase it with a third wavelength or the application of heat.

Holographic interferometry has begun to have substantial application in engineering. If a hologram of an object is made and subsequently compared with the real object, any change in the object can be seen as an interference pattern spread over the surface. Thus the elastic modes of vibration of complex objects can be determined readily. This technique is already in routine use, for example, as a production-line test method in the fabrication of automobile tires (see Figure 4.48). The number of applications of lasers and laser technology is legion and growing rapidly. At the moment, the only limitations appear to be the imagination of the user.

It would be wrong, however, to suggest that contemporary optics is concerned only or even primarily with lasers. Optical physicists are still very much concerned with atomic and molecular spectroscopy and solid-state optical phenomena. In much of this activity, the useful devices that are anticipated motivate and justify financing the work, but the research involved is frequently of the most fundamental character.

One of the older but still highly productive branches of modern optics is that concerned with thin films (Plate 4.XI). It is now possible to design and construct multilayered thin films that will reflect chosen wavelengths and transmit others in complete analogy to the much more familiar electrical filter systems. These may find very important applications in the large-scale utilization of solar power in which they can be designed to

PLATE 4.X Stop-motion photograph of ultrashort pulse [green in color; see original in *Physics in Perspective, Volume I*] in flight through a water cell. The pulse was moving from right to left. The scale is in millimeters. The camera shutter opening time was about 10 psec. The spot [red in color] is the impression left on the high-speed Ektachrome film by an infrared laser pulse used to activate the ultrafast shutter. A neodymium glass laser generates the infrared pulse, which in turn gives rise to the green pulse by passing through a nonlinear optic crystal (not shown). This photograph is thus simultaneously an illustration of second harmonic generation from the infrared ([red] spot) to the green [pulse]. [Source: M. A. Duguay, Bell Telephone Laboratories.]

PLATE 4.XI A thin-film prism. A ZnS film was deposited on a glass substrate. The film in the triangular area is thicker than that in the rest of the area. A helium–neon laser light propagating in the film was deflected after passing through the triangular area that acted as a prism. [Source: P. K. Tien and R. Ulrich, *Journal of the Optical Society of America, 60*, 1325 (1970).]

NONDESTRUCTIVE TESTING

hologram recording

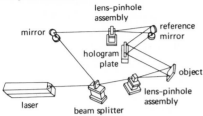
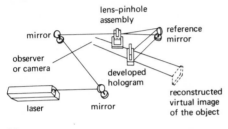

(a)

hologram viewing

(b)

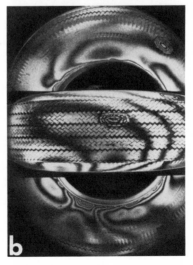

FIGURE 4.48 Nondestructive testing. *Upper left:* Beam-path scheme for the laser hologram process. (a) Construction of a hologram diffraction pattern in a photographic emulsion. (b) Reconstruction of the three-dimensional image in space.

Right: Testing of automobile tires. (a) Holographic tire analyzer reveals (b) hidden tire defects, such as tread- and shoulder-area ply separations, when the tire is inflated or subjected to various temperatures, captured here by double-exposure interferometry. Real-time viewing of a defect may be likened to observing the telltale moving ocean-wave patterns in the vicinity of a sunken rock. (GC Optronics, Inc.)

[Source: J. R. Zurbrick, *Yearbook of Science and Technology 1971.* Copyright © 1971, McGraw-Hill Book Company, Inc. Used with permission of McGraw-Hill Book Company.]

inhibit reradiation of the absorbed energy in the form of infrared radiation as described earlier in this section. High-pass, low-pass, bandpass, or low-band filters can be built economically and reliably. An interesting class of such filters transmits only a narrow band of wavelengths; when made with one or more layers tapered in thickness, the filter will transmit different wavelengths in different regions. Such wedge filters can be used to build a particularly simple and compact spectrometer for use in space vehicles or in inexpensive commercial instruments.

Another important development of thin-film technology is the high-efficiency mirror, reflecting 99.6 percent to 99.8 percent of the incident light in a chosen wavelength region. The availability of such mirrors has been one of the indispensable tools of the technical development of the laser. Still another by-product is the well-known low-reflection coating. Without such coating, complex microscope or camera lenses produce images having only low contrast because of the veiling glare of multiple reflected light inside the barrel of the lens. With such coating, complex high-performance lenses become possible.

Optical thin films have also played an important role in military infrared devices by making them wavelength-selective so that military targets can be distinguished from background sunlight. Progress in the design of very-high-precision optical systems, using large-scale computer techniques, has reached its highest point in the development of such remarkable systems as those used in the U-2 high-altitude planes and surveillance satellites. The possibility of such surveillance is fundamental to current discussion of, and hope for, international disarmament activities.

One of the most striking new developments in optics is the use of large-scale digital computer techniques for image enhancement in such fields as clinical x-ray fluoroscopy and the analysis of the Mariner photographs of Mars (see Figure 4.49). Not only can substantial improvement be achieved in apparent signal-to-noise ratios in such photographs; but also, the final images can be corrected for either deficiencies in the original optical system or lack of adequate focusing during the original exposures. The apparent improvements are frequently startling. The development of practical, mass-produced image-intensifier systems has dramatically reduced the radiation exposure required in medical fluoroscopy and has pushed back the observable frontiers of our universe when used with large optical telescopes. Paralleling this conquest of the ultralarge is that of the ultrasmall. Within the past year Crewe and his associates have succeeded in photographing the outlines of the double-helix structure of the DNA molecule through judicious tagging with thorium atoms.

Indeed, the scanning electron microscope, in itself, has given biologists, metallurgists, and all who deal with microscopic phenomena a perspective

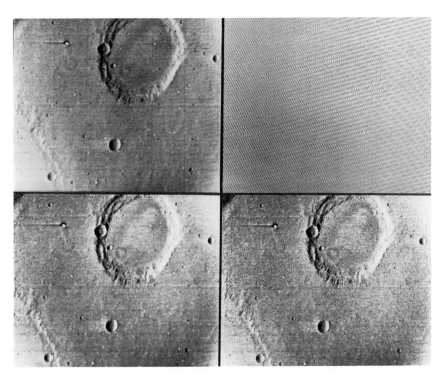

FIGURE 4.49   Example of preliminary computer enhancement of Mariner 6 near-encounter picture 18 of Mars. A portion of this frame, as originally recorded, is shown at the upper left. Apparent in it is a faint basket-weave pattern due to electronic pickup in the sensitive preamplifier of the camera system, as well as a general softness due to the limited resolution of the vidicon image tube. Computer analysis reveals the pickup pattern shown at the upper right. When the appropriate numerical value, determined from this pattern, is subtracted from each of the 658,240 elements of the picture, the result is as shown at the lower left. Two further computer programs may then be used to compensate for the smearing due to the vidicon tube, with the result shown at the lower right. The final processing procedure will involve more refined versions of these steps, as well as programs designed to remove the numerous small blemishes, correct for electronic and optical distortions of the image, and correct for the photometric sensitivity of each picture element of the vidicon tube. The computer will also be used to combine the digital and analog video data into a single, photometrically accurate picture. [Source: *Yearbook of Science and Technology 1971.* Copyright © 1971, by McGraw-Hill Book Company, Inc. Used with permission of McGraw-Hill Book Company.]

and depth of focus that, in effect, offer an entirely new window on their subjects (see Figure 4.50). The contrast between the visual impact and the immediate information content of a transmission electron micrograph

[151]

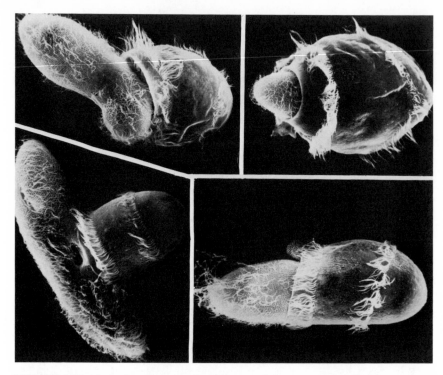

FIGURE 4.50 A single-celled animal captures and devours another in this sequence of pictures taken with the aid of the electron microscope. Exceptional details in the drama indicate the potential of this instrument. [Source: This photograph was provided through the courtesy of G. Antipa and H. Wessenberg, Argonne National Laboratory, and the *Journal of Protozoology*.]

as recently as a few years ago and what is routinely available now from even relatively inexpensive scanning electron microscopes is remarkable. It is worth noting, too, that a significant fraction of the new technology underlying these new instruments—ultraprecise magnetic lenses, superconducting lenses, and the like—has been derived from frontier projects in subfields as diverse as elementary-particle and condensed-matter physics.

Physiological Optics

Physiological optics has always been an important part of optical science, and physicists still have a role in it, although it is secondary to the work of electrical engineers and psychologists. Observations of the eyes of animals and humans enhance the understanding of the scientific basis of pattern recognition. Mechanisms have been found in the eye of the frog that respond to certain moving patterns and not to others. The human eye loses its response if an image is held stationary on the retina, showing that tremor is a necessary condition, not a defect, of vision. New adaptive mechanisms have been found in the eye to assist in the perception of color and contrast. That some parts of the pattern-recognition process are visual instinct in nature and some computed in the brain is now recognized. The human eye remains one of the truly remarkable optical instruments in its sensitivity, range, and efficiency. The processes of human vision retain large areas of mystery. Research in these areas is of direct physiological value and affords possibilities of guiding certain optics technology.

Applications of Optics in Other Sciences

Because observations are at the heart of all experimental science, optics plays a crucial part in research in nearly all disciplines. The range of optical instruments and techniques involved in physics beggars description. They are so widespread and commonplace that their origin in optics is frequently entirely forgotten. However, there are many specific and specialized applications.

The use of optical techniques in astrophysics and astronomy is obvious and requires little comment. New techniques of image enhancement, image digitization, aperture synthesis, and long-baseline interferometry have only very recently revolutionized broad areas of research in these fields.

When light from a laser beam is scattered by high-energy electrons from an accelerator, the scattered light has a higher frequency because of the Doppler effect. If the electrons have very high energies, the scattered photons become nearly monochromatic gamma rays, with energy in the range of several billion electron volts, and retain the polarization of the original light. The polarized gamma-ray beam obtained in this way is nearly free of low-energy background radiation and is ideal for many studies in particle physics.

Optical monitoring of the earth's atmosphere already has become an important tool for weather prediction with now-familiar global cloud cover satellite pictures—pictures that even a few years ago would have been considered little short of miraculous. In addition, and perhaps more important, the properties of the upper atmosphere can be measured and the effects of pollutants on the lower atmosphere can be determined from the satellite and rocketborne spectral observations of various atoms and molecules. Satellite observational techniques, using ultraviolet, visible, and infrared spectroscopy, make it possible to monitor continuously the global distribution and vertical profiles of natural and man-made variations of specific chemicals in the earth's atmosphere and to have rapid knowledge of changes. These techniques provide an immediate means by which a physicist can apply his specialized knowledge to problems of pollution and ecology. They also have provided a new observational basis for earth and planetary science.

Recent developments in frequency synthesis from the microwave to the optical region make it feasible to consider defining length and time with the same molecular transition. These achievements, coupled with the stabilities obtained through coherent laser stabilization by saturated absorption of molecules, probably will lead to the development of new and more-precise standards. Frequency synthesis to a CO_2 laser transition near 10.6 μm and stabilization of this transition by saturated absorption in SF_6 have already been demonstrated. The application of similar techniques to the near infrared is anticipated in the next few months, and following that, their extension to visible frequencies.

The newly developed saturation absorption (Lamb dip) spectroscopy can attain spectral resolution exceeding one part in 10^{10}. Thus an additional powerful tool is available for studying spectral line shapes and level splittings to within natural line widths of atoms and molecules. The new super-Kerr cell systems (see Figure 4.51), developed by Duguay and his collaborators, with effective shutter times of picoseconds, will make possible the detailed following of atomic and molecular interactions in a totally new fashion. These phenomena will be widely exploited in atomic and molecular physics and in chemistry.

There are many other applications of optics. Of particular interest recently are the laser measurements that will show variations in the earth–moon distance during the next few years and should offer a new and highly accurate check of gravitational theory. These data also will improve substantially present knowledge of: (a) lunar size and moment of inertia, (b) the earth's rotation rate and polar rotation, and (c) the present rate of continental drift. Uncertainties of less than 6 in. are involved. Further, satellite-based optical navigational systems hold promise of unprecedented precision, measured in inches anywhere on the earth.

Future Opportunities in Optics

Optical scientists and optical techniques can make major contributions in a variety of fields in addition to those already described. Consider, for example, the wretched state of technology in the Postal Service compared to that in color television or the telephone system. Or consider the fragmented leadership and lack of innovation in crime prevention and detection. A rather trivial optical system, introduced a few years ago to keep

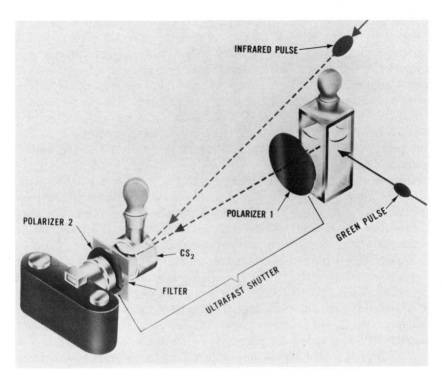

FIGURE 4.51 Setup used to photograph a green laser pulse in flight through a water cell. A drop of milk was added to the water to enhance light scattering and thus render the pulse brightly visible from the side. The ultrafast shutter is essentially an electrodeless Kerr cell. A powerful ultrashort infrared laser pulse is used to induce Kerr birefringence in the liquid, thus opening the shutter for ~10 psec. The filter greatly attenuates the infrared pulse and prevents it from damaging the camera. Both pulses are generated simultaneously by a Nd:glass laser. [Photograph courtesy of M. A. Duguay, Bell Telephone Laboratories.]

track of at least some of the nation's freight cars, excited wide newspaper coverage and attention. A great stimulus to research and development, and thus to the effective use of talented manpower, will arise as federal agencies establish programs and laboratories of their own commensurate with the complexity of their tasks and the importance to society of their missions.

The development of the picturephone is progressing and will require substantial optical support. Optical communication through glass fibers is beginning to be taken seriously, and useful links probably will be installed experimentally within two years. These two efforts could stimulate one another. Further, the picturephone will call for a wide variety of peripheral devices to introduce signals into it and distribute the output.

The importance of the picturephone concept can scarcely be over-emphasized in the context of the congested traffic and travel systems. Picturephone linkages of high quality, and with appropriate privacy controls where required, can reduce substantially the amount of business travel now considered essential. A linkage such as that now in experimental use between the Bell Telephone Laboratories at Murray Hill and at Holmdel permits effective large or small conferences, with participants at both laboratories fully involved both technically and psychologically. This mechanism may offer a partial solution to the national air and surface traffic problems.

Optical memories for computers are still in the developmental stage, and their utility is not yet fully demonstrated; but the need is so great and there are so many possibilities in optical techniques that substantial additional work will undoubtedly take place. In particular, the costs and the relationship between capacity and real-time appear favorable.

Further in the future lies integrated optics; this combines a laser beam trapped in a surface film, which is modulated or deflected by electrical or acoustical signals. In such devices, it is now possible to construct literally thousands of gate and switching elements in optical paths of fractional millimeter length and to operate them simultaneously in parallel with beams of different frequencies from integrally constructed lasers. The possibilities for miniaturization of digital instrumentation through this approach are enormous.

Much emphasis today is placed on optical methods of detecting and measuring atmospheric contaminants. Undoubtedly, there will be new problems and jobs in this field, but it seems unlikely that they will be sufficient to justify more effort than already is being allocated to this branch of physics.

Fiber optics, which for a few years was both challenging and lucrative, is now a mature and highly competitive industry with many groups ready to take advantage of any new applications that are identified.

Pattern recognition seems to hold great promise. It is possible that pattern recognition in the next ten years could change society in fundamental ways at the same time that it changes the understanding of the ways man gains knowledge and interacts with his environment. Much of the drudgery in which man engages in a highly industrialized society consists of optical pattern recognition of some kind, from the checkout counter in the supermarket to the reading of electrocardiograms. Physicists have an opportunity to make a fundamental contribution to this growing field.

It is clear that optics will continue to be of great value in military problems. Aerial reconnaissance and the image-intensifier telescope have proven invaluable to American troops in Vietnam. Reconnaissance is not only a tool for waging war more effectively but a tool for maintaining peace. It is one of the few techniques available for verifying compliance with international agreements such as arms limitations.

Structure of U.S. Optics Activity

The applied science character of optics led to a steady decrease in emphasis on optics in the physics departments of U.S. universities until about five years ago when the influence of the laser began to be felt. Through the same period, however, the Optical Society of America continued to grow, as more and more people found challenging opportunities in optics. There were some 6000 members of this Society in 1970. In 1962, the Society establishd a new journal, *Applied Optics,* which now publishes 50 percent more pages annually than does the *Journal of the Optical Society of America,* the traditional journal of this subfield. Optics thrives today primarily through its applications, though it also has strong roots in fundamental physics.

Who are the optical scientists of today? The 1970 National Register of Scientific and Technical Personnel shows 3280 physicists who indicated optics as the specialty in which they were employed. Of these 3280, one third (1111) held a PhD degree and accounted for 6.7 percent of the total PhD population in physics. The number of PhD's in this subfield shows a substantial increase from the 743 included in the 1968 Register survey, a growth rate of about 25 percent per year at a time when the rate of increase of PhD's for physics as a whole was about 7 percent per year. (See Figure 4.52).

Of the physics participants in the 1970 Register survey, 2494 indicated membership in the Optical Society of America. Other disciplines with substantial representation in the Society are psychology, chemistry, and electrical engineering. In addition, many of the members are technicians and manufacturers of optical products.

Because many practical devices are optics-limited, strong optics groups engaged in applied research and design and development work have been established in many industrial laboratories. Half of the optics PhD's work in industry. This pattern is well illustrated by the development of the ruby laser at Hughes Aircraft, the gas laser at the Bell Telephone Laboratories, and the glass laser at American Optical Company. And this pattern continues. Exceptional work in ultrashort light pulses is taking place at United Aircraft, IBM's Thomas J. Watson Research Center, and Bell Telephone Laboratories. Liquid crystals were developed at Westinghouse, and stabilized-frequency lasers at Perkin-Elmer.

One fifth of the optical scientists are working for the U.S. Government

[157]

OPTICS

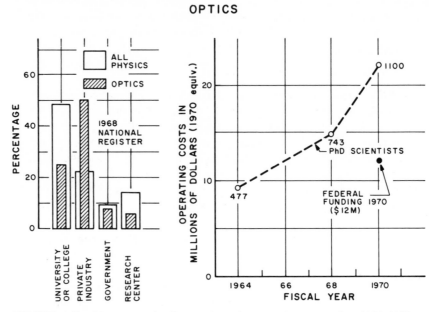

FIGURE 4.52   Manpower, funding, and employment data on optics, 1964–1970.

or in federally funded research laboratories. Since many of the laboratories were established only recently, they probably represent a substantial fraction of the new jobs in optics in the last decade. In addition, the equipment needs and extramural research programs of these laboratories have stimulated industry to employ many optical specialists (70 percent of all optical scientists work in industry).

A remarkable development of optical skill has occurred in government-sponsored nonprofit laboratories, such as the Cornell Aeronautical Laboratory, University of Michigan Institute of Science and Technology at Willow Run, and Lincoln Laboratories at MIT. Most of these groups are concerned principally with the development and operation of large special-purpose optical equipment, but many of them conduct significant fundamental studies of the underlying physics. Some have made major contributions, such as the development of the hologram by Leith and Upatnieks at Michigan.

Finally, a smaller part of modern optics has developed in university laboratories. For example, nonlinear optics was discovered at Michigan and has been studied with notable skill by Bloembergen at Harvard. But, although a very large fraction of contemporary optical scientists were trained in physics, optics has received relatively little emphasis in most physics departments. Most scientists entering this subfield have been trained primarily in some other branch of physics. However, this situation is changing rapidly. One fourth of the optics PhD's now work in academic institutions.

The hidden character of optics support makes it very difficult to establish its magnitude with any degree of certainty. Basic research in optics as a branch of physics has had very little *direct* government support in the past and seems unlikely to obtain it in the future. The estimate of 3 percent of the total federal funds for basic research in physics, in 1970, shown in Figure 4.53 may be in error.

Problems in the Subfield

Optics now draws its support primarily from industry and from government mission-oriented agencies. Both sources have invested substantial sums in basic research in optics. This pattern of support should be continued and widened. There are optical needs in transportation, the Postal Service, crime control, environmental control, printing and publishing, training devices, mapping and surveying, earthquake prediction, and information

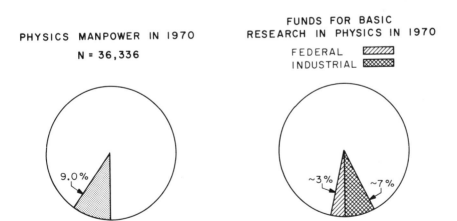

FIGURE 4.53   Manpower and funding in optics in 1970.

[159]

storage and retrieval that await exploitation. Many of these needs can be funded only at the national level. Federal agencies should recognize their opportunities and responsibilities for the support of research in optics.

Hearing may be devided into Direct, Refracted and Reflex'd, which are yet nameless unless we call them Acousticks, Diacousticks, Catacousticks, or (in another sense, but with as good purpose), Phonicks, Diaphonicks, Cataphonicks.
                    RT. REV. NARCISSUS, Lord Bishop of Ferns and Leighton
                    Philosophical Transactions of the Royal Society (London), 14, 473 (1683)

ACOUSTICS

Introduction

To most, the term, acoustics covers the process of hearing as well as those devices, methods, and materials that assist the hearing process, for example, hearing aids, microphones, loudspeakers, acoustical materials, and room acoustics, but no more. The average physicist might add the study of speech, high-frequency sound (ultrasound), and noise. However, a perusal of the pages of the *Journal of the Acoustical Society of America* will convince the reader that all elastic and inelastic waves in matter come within the realm of modern acoustics. Therefore, music, architectural acoustics, shock waves and sonic booms, vibrational phenomena, high-intensity sound, and seismic waves are all part of acoustics. The problem in this brief review is to group these manifold phenomena into a brief presentation that indicates how such a wealth of subject matter has developed.

Acoustics can be defined most succinctly as the study and use of mechanical waves in aggregate matter. The waves need not be elastic; they can be inelastic, with attenuation and dispersion, and can have large amplitudes, with consequent nonlinear effects. Acoustics deals in large part with techniques and devices for the generation, transmission, and detection of these waves. The study of the transmission of acoustic waves in matter is one of three main techniques (acoustic, electromagnetic, and the interaction of matter with particles) used to investigate the properties of matter.

Because acoustics is so old a subject, early having been associated with a basic sensory and perceptual mechanism, it sometimes is viewed only as

[160]

a part of classical physics and as a subject fully explored in the nineteenth century, which is not true. Lord Rayleigh's famous *Theory of Sound,* first published in 1877, did not represent the end of acoustics. On the contrary, the theory and Rayleigh's subsequent research contained the elements of a new beginning in which new concepts and techniques were generated. These new approaches and methods not only caused but resulted from new applications in science and technology. Consequently, acoustics has today both a classical and a quantum character.

As with optics, and more generally with electromagnetic radiation, acoustics has a wide spectrum ranging from approximately 10^{-1} to 10^{14} Hz, as indicated in Table 4.4. Generally, the techniques differ greatly in the different ranges, although the physical principles are similar.

Acoustic waves often penetrate media (for example, oceans or solids) that electromagnetic waves do not enter. For example, all our knowledge about the interior of the earth (liquid core, temperatures, and pressures) comes from acoustical studies together with equations of state, which are determined largely with acoustical measurements. This example is but one of many in which acoustics provides the only means available for direct study of the properties of matter.

Defining the scope of acoustics is difficult, for the subject matter of this subfield cuts across the definitions of the subfields of physics selected by the Physics Survey Committee. In fact, acoustics has extensive ramifications not only in the various subfields of physics but in many other scientific disciplines and in technology. Consequently, people using the concepts and techniques of acoustics can be found in many physics subfields, related sciences, and engineering disciplines. However, in spite of the variety of topics dealt with in acoustics and the strong association with other sciences, engineering disciplines, and technologies, acoustics remains fundamentally a part of physics. It includes all material media. It requires the mathematics of theoretical physics. Its methods play a primary role in exploring the characteristics of the various states of matter. In addition, it provides

TABLE 4.4 Frequency Ranges in Acoustics

Frequency (Hz)	Name of Frequency Range
10^{-1}–20	Infrasonic
20–2×10^{1}	Audio
2×10^{1}–5×10^{8}	Ultrasonic
5×10^{8}–10^{12}	Hypersonic or praetersonic
10^{12}–10^{14}	Thermal vibrations

many techniques, devices, and inventions for other subfields. And, in education, there is a unity about the subject of acoustics and its manifold applications that tends to be lost if it is taught in a fragmented way in a number of different courses.

The constantly changing nature of acoustics as new discoveries occur also contributes to the problem of developing a practical definition of this subfield. Subjects within acoustics grow and gradually separate from it as understanding progresses through the development of the geometry of transmission to the response of materials to acoustic waves. Thus, acoustics stimulates the development of new areas of investigation that later become either independent or subdivisions of other physics subfields that deal with the properties of matter. However, the techniques of generating and receiving waves, as well as the geometrical aspects (as in geometrical optics) of transmission, remain a fundamental part of acoustics.

The currently changing emphases in physics also have an impact on the definition of acoustics. Emphasis on more applied work relevant to critical national problems necessarily means increased attention to acoustical studies. Examples of disciplines that are undergoing major development with acoustical methods are biophysics and geophysics. The large programs planned for oceanography and ocean engineering will be limited by the available understanding of acoustics.

In addition, the shift in emphasis of federal programs from defense to environmental and social problems (pollution, public transportation, housing, health and safety, and communications) will have an impact on the relative place of acoustics in the hierarchy of physics subfields.

In public transportation, there are critical problems of noise, shock waves, materials, communications, and control. Physics, and acoustics in particular, has much to contribute to their solution.

Recent developments in acoustical holography promise to make striking contributions to problems of health. One example is shown in Figure 4.54. Further, safety devices necessarily involve electromagnetic and acoustical signals. Electromagnetic–acoustic devices now exist that will locate astronauts to within 1 ft of a spacecraft at all times. The potential applications of such devices have not yet been fully explored.

The recent developments of ultrasonic surface-wave physics will probably exert a strong impact on the communications and computer industries, which by now pervade all aspects of modern life with examples familiar to everyone.

In this brief summary, acoustics is discussed under the headings of instrumentation and the relation of acoustics to the medium, to man, and to society.

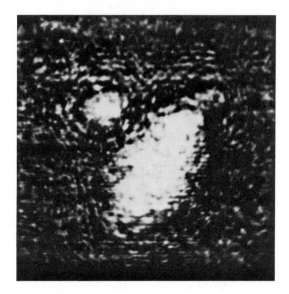

FIGURE 4.54 Reconstruction with visible light of acoustic hologram of a living human eye with a retinal detachment. (P. Greguss, Jr., Budapest Medical University.) [Source: *Yearbook of Science and Technology 1970*. Copyright © 1970, by McGraw-Hill Book Company, Inc. Used with permission of McGraw-Hill Book Company.]

Instrumentation

Seventy-five years ago the sound sources available to man were limited largely to the human voice, musical instruments, and a few whistles and other mechanical devices, all producing sound energy almost exclusively in the range audible to the human ear—20 to 20,000 Hz. The discovery of the piezoelectric effect by Jacques and Pierre Curie in 1880, the effect in which mechanical stresses on many crystalline solids can produce electrical voltages and *vice versa,* subsequently was used widely to produce ultrasonic waves (ultrasound) whose frequencies can be as much as a million times greater than the highest frequency audible to the human ear. At the other extreme of frequency, the use of geophysical receivers (geophones) and of low-frequency (infrasound) detectors has led to an entirely new range of measurements. With a modern infrasound detector, for example, it is possible for a detector in New York City to hear an Apollo launch at Cape Kennedy. The range of acoustic measurements now extends from below 1 cycle per hour to well over 5×10^9 Hz.

The individual techniques for the production of ultrasound are myriad. The discovery, during World War II, that certain ceramic materials, such as barium titanate, could be electrically polarized and used in a fashion similar to piezoelectric crystals made available electromechanical transducers that could be molded into virtually any shape or size. Another technique involves the generation of ultrasound by means of semiconducting films, such as those of cadmium sulfide, painted onto the test material. Most recently, the direct production of ultrasound through the impinging of electromagnetic radiation on a metal surface has become a practical method for the generation of high-frequency sound in solids.

Acoustics and the Medium

Although the mathematical analyses of the processes by which sound is absorbed in its passage through air and other media were well developed in the nineteenth century, accurate experimental measurements were lacking even 50 years ago. The reason is not difficult to find. In the range of audible frequencies, the absorption is so small that enormous path lengths would be needed for measurement—distances that were simply too great to be experimentally feasible. Furthermore, the wavelengths of such sounds are so long that diffraction effects enormously complicate experimental measurements. It was not until substantially collimated beams of ultrasound became available that reliable measurements on acoustic propagation were obtained—between 1920 and 1940 for gases, 1930 and 1950 for liquids, and more recently for solids.

These measurements produced a major surprise. In air, the absorption coefficient, the measure of energy loss in the medium, was considerably greater than that predicted through consideration of the viscosity and thermodynamic conditions developed by Stokes and Kirchhoff in the mid-nineteenth century. The corresponding measurements in liquids yielded values for the absorption coefficients that were often 1000 or even 10,000 times the predicted ones.

These findings quite naturally led to vigorous activity, both theoretical and experimental. The result was the identification of relaxation processes, involving the excitation of internal rotational and vibrational states of the molecules involved, as the cause of this extra absorption in gases. In liquids, the relaxations frequently could be associated with structural rearrangements inside the molecule or aggregates of the molecules involved. These studies resulted in a steady accumulation of acoustical information about liquids and the successful use of this information as a probe to study the structure of the substances involved. (See Figure 4.55)

Because sound absorption generally decreases as the medium becomes more condensed, absorption measurements in solids at low frequency are even more difficult than in liquids; therefore, it is not surprising that absorption measurements in solids are the most recent. Measurement of sound velocities and absorption in different directions in single crystals and

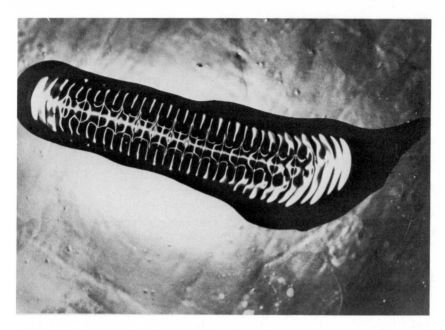

FIGURE 4.55 Visible effect of sound on a layer of glycerin at 30 Hz and weak amplitude. (Photo by J. C. Stuten, in H. Jenny, *Cymatics,* Basilius Presse, 1967.) [Source: *Yearbook of Science and Technology 1971.* Copyright © 1971, by McGraw-Hill Book Company, Inc. Used with permission of McGraw-Hill Book Company.]

in crystals under various external stimuli, such as magnetic fields, varying pressures, or varying temperatures, have contributed significantly to the understanding of crystalline imperfections, the behavior of superconductors, and the nature of Fermi surface metals.

A number of areas in physical acoustics merit special mention because of their recent rapid growth, for example, surface waves, acoustic holography, high-accuracy velocity-change measurements, nonlinear acoustics, acoustic emission and heat pulses, and infrasonics.

As a group, these research areas typify the steady evolution of investigations in physical acoustics from basic physical research studies to applications in other sciences and in technology. Two examples are given in Figures 4.56 and 4.57.

Surface waves, known from the days of Lord Rayleigh, only recently have been exploited for their applications to electronic systems. They have marked advantages over bulk waves in information storage and in signal filtering. They require only a single surface and are better suited to piezo-

[165]

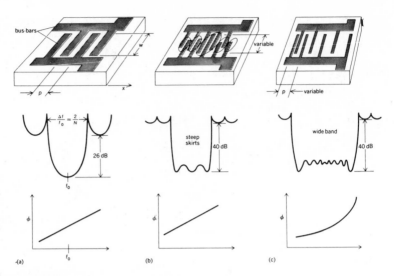

FIGURE 4.56 Transducer configurations and corresponding insertion loss and phase characteristics of electric filters: (a) interdigital construction; (b) amplitude weighting technique; (c) graded periodicity technique. [Source: *Yearbook of Science and Technology 1971*. Copyright © 1971, by McGraw-Hill Book Company, Inc. Used with permission of McGraw-Hill Book Company.]

electric amplification because of their low surface-to-volume ratio. As filters, they allow the designer to prescribe the phase characteristic independent of the amplitude characteristic.

Acoustic holography, through development of a water–air interface method of acoustical imaging, allows the transfer of the spatial modulation of a sound field onto a light field, which can then be used to produce optical images. At the water–air interface, the surface of the water is deformed by the incident sound pressures. A light beam is then reflected from this deformed water surface to obtain the required spatial modulation. Acoustic holography offers new possibilities for more accurate imaging of objects within the human body and for nondestructive-testing applications generally.

The use of the ultrasonic imaging technique in nondestructive testing is increasing rapidly. This technique is useful, generally, for detecting the same types of flaw as those recommended for conventional ultrasonics tests. In the medical field it may find use in studying the movement of fluids in the body. An ultrasonic image of a fingertip is shown in Figure 4.58.

High-accuracy measurement of velocity change has proved useful in the study of solids at low temperatures. With recently achieved increased

[166]

FIGURE 4.57 A Zenith-engineered thick-film circuit (foreground) is designed to replace the present-day large television intermediate-frequency amplifier in background. Its tiny ultrasonic filters (at pencil tip and left) select certain frequencies and reject others. [Photograph courtesy of Zenith Radio Corporation.]

sensitivity, these measurements are now used in the determination of third-order elastic constants, especially of soft materials. These constants have application in many aspects of the solid state, for example, establishing of equations of state, providing connections between measured mechanical properties and thermal effects (through measurement of the Grüneisen constant), and developing an acoustic thermometer.

The work on third-order elastic constants in solids was preceded by a burst of corresponding nonlinear acoustical research in liquids. This led to the determination of higher-order coefficients in the adiabatic equation of state for liquids, and also to the realization and study of the sound-with-sound interaction and the practical application of high-resolution parametric sonar.

The discovery of emission from a single dislocation wall in a solid has

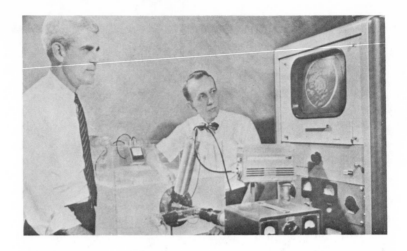

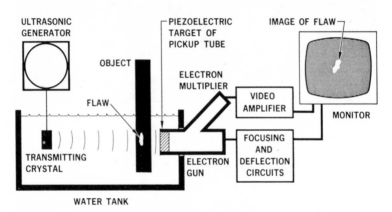

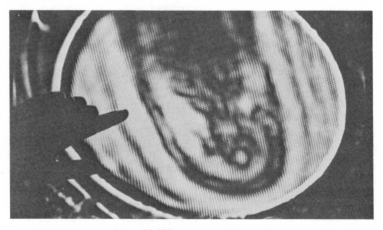

[168]

◄ FIGURE 4.58 *Top:* Photograph and diagram of an ultrasonic inspection system that yields a televised picture of the ultrasonic radiation reaching the pickup tube detector. Water is used to transmit the ultrasonic vibrations in a flashlight-type beam from the transmitting crystal (left) to the object and then to the detector. This detector is a camera in a closed-circuit television system. (Courtesy James Electronics, Inc.)

Bottom: Television picture of a human finger obtained with the ultrasonic system shown above. The image of the finger is the oval in the center of the round white area, which is the image of the sensitive part of the television camera tube. The tip of the bone in the finger is easily seen. The hand at left points to the television image.

[Source: Atomic Energy Commission, Division of Technical Information.]

suggested the possible identification of characteristic signatures of various solid defects, and thus the creation of yet another new nondestructive-testing tool. The use of heat pulses as sources of acoustic waves in the 10^{11}- to 10^{12}-Hz range has also become possible recently, leading to new investigations in the physics of condensed matter.

Infrasound studies deal primarily with the pressure fluctuations in the atmosphere, with periods from 1 sec to several hours. Causes of such pressure fluctuations are mountain-induced vortex shedding by the jet stream in the upper atmosphere, severe storm systems, aerodynamic turbulence, and volcanic eruptions. Infrasound has defense applications in the detection and identification of major rocket launches. It also offers the possibility for major advances in the knowledge of the large-scale behavior of the atmosphere, with applications to clear-air turbulence detection, the tracking of severe weather systems, and the capability to provide advance warnings of destructive open-sea earthquake waves or tsunamis.

When research in plasma physics bloomed, it was early noted that the ionized medium was sensitive to the passage of sound waves. In addition to the normal, longitudinal sound wave, an additional transverse wave appears, which is known as the Alfvén wave, after its discoverer, the 1970 Nobel laureate in physics.

Study of the propagation of sound in liquid helium also has been productive. In the 1940's, Landau predicted the existence of a periodic temperature wave in liquid helium below the λ point. This wave, known as second sound, was later detected experimentally by his colleague, Peshkov. The study of first and second sound in liquid helium resulted in the identification of two other forms of sound. Third sound is the longitudinal oscillatory motion of the superfluid component of a thin superfluid helium film. The motion is parallel to the surface of the film and involves only the superfluid component. Fourth sound, a compressional wave of the superfluid component, moves through the pores of a finely dispersed solid when the pores have been filled with liquid helium. The study of all these phenomena provides an important method for probing the physical character and quantum behavior of liquid helium.

In geophysics and ocean science, the respective media, the earth and the ocean, provide gigantic laboratories in which ultrasonic, sonic, and infrasonic waves are used routinely to provide data that frequently are inaccessible by other means.

Study of the ocean requires a detailed knowledge of ocean depths and the nature of the ocean floor. Information on both can be obtained from a study of sound transmission and scattering. The problems of underwater communication and sonar development have led to sophisticated use of computer, correlation, and filter techniques to detect minute signals in a world of noise.

In recent years, the observed periods and amplitudes of the fundamental modes of vibration of the earth have enabled scientists to derive a more accurate model of the earth's interior. The implanting of a geophone on the surface of the moon has produced similar data regarding the moon's interior. By the use of suitable systems of recording instruments, relative changes in the earth's internal displacements can also be monitored, so that we now have a realistic possibility of achieving earthquake predictions—a feat that could save untold thousands of human lives, quite apart from the material advantages to be gained from such predictions.

Acoustics and Man

Speech It has long been a goal of speech scientists to produce synthetic speech. The development of sound recording and reproduction techniques already has been combined with the computer storage and retrieval of information to such an extent that a breakthrough appears imminent. A major problem has been the modifying effect that adjacent sounds (phonemes) and near-adjacent ones have on any given speech sound. The basic question is: How does one program a computer so that it will select, modify, and present the sequence of sound that carries the desired meaning?

Hearing The process of hearing involves both psychological and physiological acoustics. Two recent, fundamental discoveries have been made in psychological acoustics. The first is that man apparently possesses an internal psychological scale that enables him to express the experienced auditory sensation magnitudes quantitatively, with reasonable consistency, even when intersensory comparisons such as the loudness of a sound with the brightness of a light are involved. The second discovery is that the problem of auditory signal recognition reduces to a discrimination between noise-plus-signal and noise-alone events. These discoveries have led to increased social and industrial applications and to increased rigor of psycho-acoustic experimentation. For example, the instrumentation and method-ology developed for acoustic impedance measurements at the eardrum have recently been accepted for diagnosis of the highly prevalent middle-ear disorders. An example of such instrumentation is given in Figure 4.59.

An exact knowledge of the working of the human auditory system still

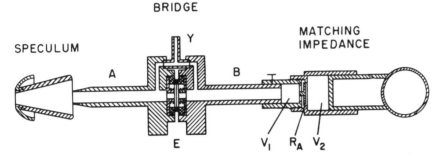

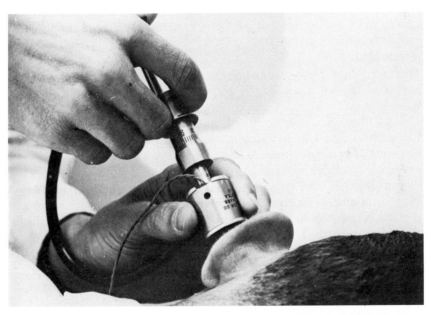

FIGURE 4.59 Acoustic bridge. *Top:* For measurement of acoustic impedance at the eardrum. Air volume V_1 is adjusted to match the residual air volume in the ear canal. R_A and V_2 are variable impedance elements. [Source: J. Speech Hearing Res., 6, 304–314 (1963).]

Bottom: Acoustic bridge held in the ear by hand. [Source: American Speech and Hearing Association, ASHA Monograph 15, Washington, D.C. (1970).]

[Photographs courtesy of J. Zwislocki, Syracuse University.]

[171]

is lacking. Extensive research is required. Among the goals of this research are the determination of the way in which sound parameters are encoded in the nervous system and the explanation of auditory characteristics—that is, the nature of the sensory brain function.

Acoustic physicists have rapidly adapted techniques from other subfields of physics to their specific purposes. As an example, the nuclear Mössbauer effect has recently been used to measure precisely the normal motions of the human eardrum and the components of the auditory mechanism through appropriate deposition of a radioactive emitter on the drum or bones. These measurements of velocities of the order of 0.1 mm/sec, or equivalently, one mile per year, confirm the remarkable fact that the threshold of human hearing corresponds to a drum motion of the order of the diameter of a hydrogen atom. Still not understood is the mechanism by which man can distinguish, with some reliability, tones differing by only a few cycles in tens of thousands!

In these areas of speech and hearing, the application of ultraminiaturized electronic components (implanted and otherwise), which are just now coming into use as a consequence of continued research in both basic and applied physics of condensed matter, undoubtedly will have revolutionary consequences in the alleviation of defects that afflict a significant fraction of mankind.

Medicine In addition to providing hearing and speech aids, tests, and the like, acoustics has a special role to play in medicine. Most of the possibilities for medical applications so far are underdeveloped or undeveloped. The use of ultrasound in therapy is an example. Many ultrasonic generators are in active use, but accurate knowledge of how ultrasonic therapy helps is still lacking. Evidence exists of acceleration of wound healing by low-intensity ultrasound and of enhanced cancer radiation therapy effectiveness through simultaneous administration of ultrasound and of x rays, but general procedures have not yet been developed. Whether it is the local heating or the modification of flow of body fluids that is the effective characteristic is not yet known.

Ultrasound has been tested widely as a probe for the visualization of internal body organs (see Figure 4.60). In this, it is similar to the x ray, although lacking many of the latter's side effects, but its use has been far more limited. Its greatest success has been in providing information on acoustic impedance discontinuities in soft tissues and in situations such as examination of the brain, abdomen, and heart, where x radiation provides little or no discrimination. The use of acoustic holography in such nondestructive testing of the human body is still in its infancy but holds promise of success.

Other techniques that have reached the clinical stage are the application of Doppler shift and returning echoes to the measurement of blood-flow

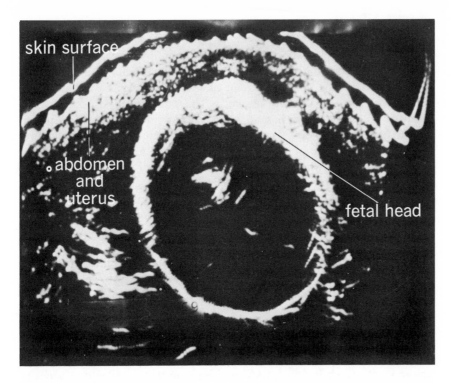

FIGURE 4.60   A typical ultrasonic scan of the pregnant abdomen at level of fetal head. Echo pattern at left may be an extremity. (University of Colorado Medical Center.) [Source: *Yearbook of Science and Technology 1970.* Copyright © 1970, by McGraw-Hill Book Company, Inc. Used with permission of McGraw-Hill Book Company.]

rates, focused ultrasonic beams in neurosurgery to cause highly localized damage deep in the brain without injuring surrounding tissues, and ultrasonic radiation of the ear to treat disorders of the balancing mechanisms.

Perhaps the medical ultrasonic application most familiar to the citizen is in modern dentistry. Use of ultrasonic drills, with their ability to cut hard dental structure without simultaneous damage to soft support tissue and without generation of excessive heat, has done much to reduce the traditional, and often irrational, horrors of the dental chair.

Acoustics and Society

Music   The impact of acoustics on society through audio recording and reproduction requires little comment. The production and reproduction of music, speech, and other sounds by means of tape and disk recorders,

[173]

pocket radios, and electronically amplified musical instruments are commonplace.

The advent of electronic and computer music, however, has opened new vistas for both composers and performers. After an initial period of bizarre experimentation, the musical content of which may well have been debatable, a new stage has been reached. The music produced by a Moog synthesizer, for example, has attracted serious interest. This instrument, typical of others as well, is capable of producing virtually any desired musical sounds, some of which resemble those of existing musical instruments and some of which are entirely new. Soon a new version of musical form and sound can be expected to evolve, and recreational composing could very well occupy the leisure time of many individuals.

The on-line computer also will play its part by allowing traditional composers to perfect their compositions with an entire orchestra effectively at their fingertips through the medium of computer storage.

Perhaps the greatest progress will be made by those trained from youth in both the musical arts and in physics, so that the novel ideas of both disciplines can be combined to produce results inconceivable to the traditional composer. Early stages of this form of education already are evident.

Architectural Acoustics   Serious musical performances depend on the nature of the room in which they are given, so that architectural acoustics must receive attention. Composers have always had some specific hall-reverberation characteristics in mind, whether consciously or not, for each of their works. Some modern composers now see the exciting possibility of the expansion of artificial reverberation to permit reverberation times that change for different parts of a composition and could be different for different musical frequencies. Here again an entirely new musical experience becomes possible.

Recent research has made it clear that reverberation time is only one of the factors contributing to acoustical quality in concert halls. Of greater importance, probably, is the detailed signature of the hall reverberation—the response during the first 200 msec after the direct sound from the orchestra is heard.

There are many other subjective attributes to musical acoustical quality besides the liveness (reverberation time), including richness of bass, loudness, brilliance, tonal blend, and related binaural spatial effects. Although the computer simulation of real halls could lead to the separation of a number of variables, it is probable that model experiments, as well as tests on full-scale halls, will be necessary to improve the basic understanding of the relative importance of the many factors involved.

[174]

Only recently have significant efforts been made to control the acoustical environment of dwellings. Building codes on all government levels will be required before major achievements can be realized. Essentially all the fundamental acoustical knowledge exists to control the transmission of sound from one apartment to another; however, new techniques of application will only develop as standards for acoustical privacy become an accepted part of building specifications. With growing encroachments on the privacy of the individual as the population burgeons, such standards are already underdeveloped.

Noise   Perhaps the principal impact of acoustics on society in the years to come will be in relation to the problem of noise pollution. In addition to the specific damage to the hearing process that can occur from excessively loud or prolonged sounds, noise that is well below the physical damage level is characteristic of contemporary urban civilization and interferes increasingly with speech, relaxation, and sleep. Control of noise will not be achieved, however, until society demands it and is prepared to pay the price in both money and possible inconvenience.

The most intense sources of noise that an average person encounters are those involving aerodynamic phenomena. The exhaust streams from jet or rocket engines, the rotating blades of aircraft propellers, and, of course, the sonic boom are all examples. It has been traditional in acoustics to remark on the small amount of power dissipated in the sounds produced by the loudest voices or musical instruments. (The analogy customarily used is that all the shouting of the fans in a large football stadium would serve only to heat one cup of tea!) Therefore, it may come as a surprise to many to learn that the power radiated as *noise* by the engines of a large commercial jet airliner would be sufficient to operate the average automobile, and that the noise power radiated by the Saturn V rocket is of the order of that required to propel a large aircraft carrier. Noise remains a big problem. It is today one of the major roadblocks to the continued orderly development of air transportation systems. Noise-control considerations and noise-reduction technology are vital in the future development of aviation (and, indeed, of mankind). Scientists from many different disciplines have made and will continue to make important contributions toward both the understanding and the solution of aerospace noise problems.

Distribution of Activity

Because of the interdisciplinary character of acoustics, research is found not only in physics departments but increasingly in engineering, geophysics, oceanography, and life science as well as music departments. In regard to

[175]

biology, most acoustics-related research is performed at universities and medical schools. In addition to university activities, government laboratories are very active in acoustics research and development. The most easily identified facilities are those operated by the U.S. Navy. These laboratories are supplemented by other national security installations, the National Bureau of Standards, and laboratories of the Environmental Science Services Administration. Industry supports wide-ranging research in this subfield.

Physicists who designated acoustics as their subfield of employment in the 1970 National Register of Scientific and Technical Personnel represented 3 percent of the 33,336 participants in this survey. The PhD physics population in acoustics represented 2 percent of the total number of PhD physicists. These data indicate that acoustics is among the least populous physics subfields and that a gradual decrease in the relative number of physicists has occurred in recent years. For example, in 1964, acoustics accounted for 6 percent of the total physics registrants.

Physicists identified with acoustics are found principally in industry; 44 percent reported employment in industry in 1970. More than one fourth (29 percent) worked in government laboratories, and 17 percent in colleges or universities (Figure 4.61). Academic employment of PhD's

ACOUSTICS

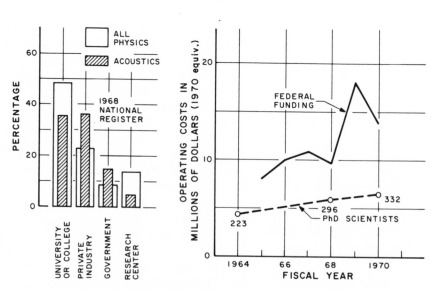

FIGURE 4.61   Manpower, funding, and employment in acoustics, 1964–1970.

was nearly four times greater than was the case for non-PhD's, with roughly equivalent percentages of doctorates indicating academic and industrial employment. Substantially higher percentages of PhD's in acoustics indicated industrial and government employment than was true of all physicists. Basic research and teaching received more emphasis among the PhD's in acoustics than the non-PhD's, but involvement in these activities was far less in acoustics than in all physics subfields taken together.

Data on the support of work in acoustics are difficult to obtain. The DOD is the major source of support for basic research in acoustics. Some relatively small-scale support comes from NASA and the NSF. The heavy concentration of acoustics personnel in industry suggests that this subfield derives much of its support from the private sector. A very rough estimate is that 3 percent of the federal funds and 1 percent of industrial funds allocated to basic research in physics in 1970 were applied to basic research in acoustics (Figure 4.62).

Problems in the Subfield

For many years, an outstanding problem facing acoustics as a subfield of physics has been the preservation of its identity. Some subject matter has been taken into other disciplines and subfields because acoustics provided a successful tool (for example, ultrasonic measurements in condensed-matter physics), some has always spread across other classifications (noise *vis à vis* turbulence), and some has been so largely outside physics that it

ACOUSTICS

PHYSICS MANPOWER IN 1970

N = 36,336

FUNDS FOR BASIC
RESEARCH IN PHYSICS IN 1970

FEDERAL
INDUSTRIAL

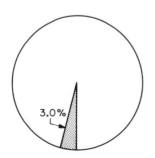
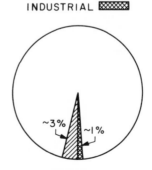

FIGURE 4.62   Manpower and funding in acoustics in 1970.

[177]

has virtually been forgotten (speech, hearing, and bioacoustics). It matters little what label attaches to the subfield, but it is important that it be maintained.

A corollary of the problem above has been instruction in acoustics and fields related to acoustics—mechanics, fluid dynamics, classical wave theory. Whether or not it is called acoustics, the study of the whole area in physics departments should be encouraged.

Finally, physical acoustics has had a long tradition of heavy support by the Office of Naval Research. The amount of this support to universities is now in the process of severe curtailment, and means to induce other agencies to support a larger fraction of basic acoustics research that is relevant to their needs must be found.

> In the next place we have to consider that there are diverse kinds of flames.
>
> PLATO (c. 428–c. 348 B.C.)
> *Timaeus* 58

> The wind goeth toward the south and turneth about into the north; it whirleth about continually and the wind returneth again according to his circuits.
>
> *Ecclesiastes* I:6

PLASMA AND FLUID PHYSICS

Introduction

The physics of fluids deals with the study of liquids and gases, and plasma physics, with the study of ionized gases. Human beings spend their lives in intimate contact with air and water, both inside and outside their bodies. The science that studies the forces and motions of liquids and gases is called fluid dynamics (the physics of fluids or fluid mechanics). Fluid dynamics covers an enormous range of topics from the most basic to the most applied. It is one of the oldest sciences in its hydraulics aspects, yet one of the newest in fluidics, the basis for a new family of computational devices. Turbulent fluids pose one of the outstanding challenges in theoretical physics; and success in the understanding and control of turbulence will have far-reaching consequences not only in all aspects of fluid dynamics but also in other sciences and technology.

When a liquid is heated, it becomes a gas. When a gas is further heated, the negative electrons are thermally stripped from the positive ions, giving

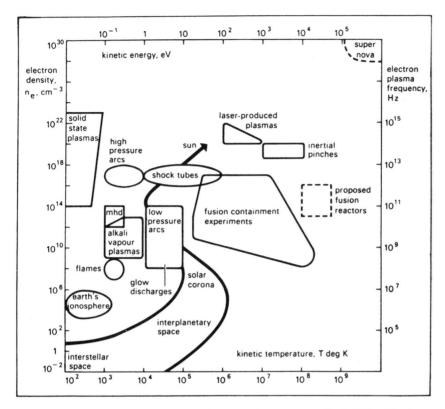

FIGURE 4.63 The plasma state—density and temperature (by I. G. Spalding).

rise to an ionized gas or plasma. This so-called fourth state of matter * exists with a very large range of physical parameters as illustrated in Figure 4.63. Particle densities range from 1–100 cm^{-3} in interstellar gas, through 10^8–10^{20} cm^{-3} in laboratory plasmas, to 10^{22}–10^{25} cm^{-3} in stellar interiors and nuclear explosions. Plasma temperatures range from fractions of an electron volt in low-current discharges to 10^5 eV in fusion plasmas and reach relativistic energies in cosmological plasmas, such as in the Crab nebula.

In its simplest form, plasma consists of fully ionized hydrogen, and at high temperatures and low densities collisional effects become negligible. This is the realm of astrophysical and thermonuclear plasmas, which has been established as an experimental science only in the last 15 years or so and which is often called the physics of fully ionized gases. In lower-temperature plasmas, the interactions among many different particle species,

* The designation, fourth state of matter, was used by Crookes in 1879; Langmuir introduced the term plasma in 1928.

positive and negative ions, molecular ions, and neutral atoms become important. In many laboratory plasmas, the interaction of these species with surfaces of condensed matter must also be taken into account. Plasma physics also incorporates many of the disciplines of magnetohydrodynamics, since in many cases plasma behaves as a conducting compressible fluid dominated by interactions with magnetic fields. In addition, studies of hypersonics—shocks and detonation waves at high Mach numbers—and high-temperature chemical flames are a part of plasma physics.

Almost all matter in the universe exists in the ionized state except for the relatively small but important fraction in planets and gas clouds. Man spends his life immersed in fluids but is surrounded on the cosmic scale by plasma.

Since so much of man's environment comprises either fluids or plasmas, it is small wonder that so much intellectual effort has been spent attempting to understand them. The most obvious and dramatic part of the environment, the weather, is the subtlest and most complicated problem in the physics of fluids. There is only one other comparable problem in fluids—comparable because there is not the slightest indication of its origin—and that is how the first giant aggregates of matter, clusters of galaxies, were formed during the early stages of an expanding universe. Between these two applications, one of great social concern and impact and the other a spectacular edifice of human reason, lies the science of fluids. An activity in plasma physics that holds high promise of great success and concomitant social impact is fusion research.

Plasma Physics and Fusion Research

Plasmas do not impinge on man's immediate senses in any fashion so direct as the weather. They exhibit a far more complex and intricate behavior than do classical fluids. This very complexity allows modes of behavior that are far removed from those normal to terrestrial matter. In particular, the extreme of ultrahigh temperature makes feasible the release of controlled fusion power on earth. This particular aspect of the behavior of plasmas—the quest for fusion power—has by and large dominated plasma physics for the last 20 years. The knowledge that has flowed from this quest has led to scientific understanding of such diverse topics in plasma physics as the behavior of the magnetosphere around the earth, of magnetic fields in the galaxy, and of the plasmas surrounding stars; the propagation of radio waves in galactic and intergalactic space; and the embryonic considerations concerning quasars and pulsars. This knowledge offers a dramatic example of how man, in the pursuit of a very practical goal, must face the overriding requirement of developing fundamental scientific knowledge.

The controlled fusion problem was not recognized initially as one requiring a great extension of the existing knowledge of plasma physics. Many technical approaches were pursued in parallel; however, grave difficulties of a plasma-physical nature seemed to frustrate them all. It gradually became clear that much more basic research would be required before any practical, large-scale device or fusion reactor would be possible. Theoretical effort directed toward the fundamental understanding of plasmas received high priority. This understanding was reinforced by many experiments in the fusion program directed more toward the development of plasma physics than toward the immediate objective of fusion power. The results were, first, steady progress toward understanding plasma physics in detail and, second, steadily accelerating progress toward controlled fusion power. A variety of experimental approaches are being pursued in the United States and abroad (see Figures 4.64 and 4.65), and technology is advancing on several fronts (see, for example, Figure 4.66). Progress has now reached the point at which realistic reactor designs and development time scales are being discussed.

At the inception of the controlled fusion project, scientists knew that they had to achieve a millionfold increase in both temperature and a somewhat

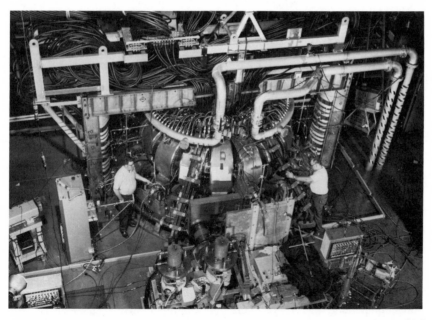

FIGURE 4.64 Controlled fusion research device of the Tokamak type (in toroidal geometry) located at Princeton University Plasma Physics Laboratory. The principal goal of this program is to measure particle and energy confinement times under many conditions and to predict how these relationships may change as the size of the torus is increased. [Courtesy Plasma Physics Laboratory, Princeton University.]

[181]

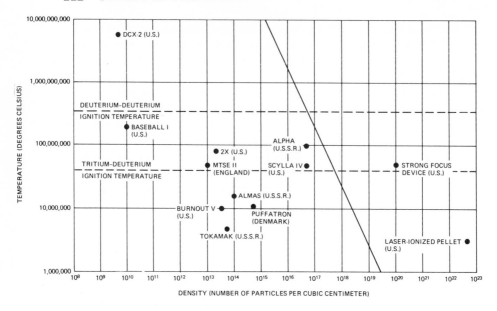

FIGURE 4.65 Plasma experiments that have achieved temperatures near or above the fusion ignition temperatures of a deuterium–tritium fuel (bottom horizontal line) and a deuterium–deuterium fuel (top horizontal line) are identified by the name of the experimental device and the country in which the experiment took place. The diagonal line represents the limit beyond which the materials used to construct the magnet coils can no longer withstand the magnetic-field pressure required to confine the plasma (assumed to be 300,000 G in this case). Beyond this limit, only fast-pulsed systems (in which the magnetic fields are generated by intense currents inside the plasma itself) or systems operating on entirely different principles (such as laser-produced, inertially confined plasmas) are possible. The record of $6 \times 10^{9\circ}$C was achieved with the aid of a high-energy ion-injection system associated with the DCX-2 device at the Oak Ridge National Laboratory. [Source: W. C. Gough and B. J. Eastland, "The Prospects of Fusion Power," Scientific American, *224*, 55 (Feb. 1971).]

abstract quantity called confinement. Confinement implies maintenance of the reacting nuclei in a restricted volume to permit the reaction to continue. At the required temperatures, no material container was conceivable, and so development of the magnetic bottle began. (A more prosaic example of confinement is the preservation of thermal energy in a thermos bottle with a tight cork.) To achieve fusion power, these two quantities, temperature and confinement, had to be increased simultaneously. The goal of past work was to obtain a 10^{12} increase in the conditions dealt with in ordinary furnaces. Remarkable progress has been achieved. At present, this factor has been reduced to about 10^3, and it is still decreasing rapidly.

[182]

Practical fusion reactors will involve either the deuterium–tritium (D–T) or the deuterium–deuterium (D–D) nuclear fusion reactions. The former is technically more feasible in the early stages of the program inasmuch as the ignition temperature for the D–T plasma is only 45 million deg, whereas that of the D–D plasma is some 400 million deg. But the former has certain disadvantages. The neutrons produced are of relatively high energy (~ 14 MeV), and these must be recycled in a lithium blanket to breed the tritium fuel in a $^6Li(n,T)^4He$ reaction. Not only will this create severe shielding problems, but also the world supply of lithium is not greatly different from that of uranium, hence ultimately limited. In the D–D case, on the other hand, an essentially limitless supply of stable deuterium is available in the oceans. Furthermore, the D–D neutrons are much lower in energy (~ 2.5 MeV) and consequently are much more readily handled. It seems clear that

FIGURE 4.66 Photograph of the 2X II pulsed magnet set prior to construction of the external glass and epoxy reinforcement. The completed structure, which contains approximately seven tons of plastic reinforced with glass cloth, is the latest and largest experimental pulsed magnet to be built using techniques that were pioneered at the Lawrence Livermore Laboratory and that are now commonplace in industrial application. [Courtesy Lawrence Livermore Laboratory.]

[183]

practical fusion reactors will be achieved via the D–T system, followed by an eventual conversion to D–D systems.

Present progress in the U.S. program suggests the possibility of a self-sustaining fusion system by 1980 and economically competitive fusion-based power sources by 2000, assuming adequate support for the program.

As noted earlier in this chapter, too, the availability of very-high-power lasers has stimulated great interest in the possibility of using such lasers to stimulate fusion reactions in a small pellet of a frozen D–T mixture. The target goal here is perhaps a tenfold increase beyond the Lawson criterion of $n\tau = 10^{11}$ needed to achieve a self-sustaining system. Here, n is the plasma density of the reacting species in particles per cm^3 and τ is the containment time in seconds. In the magnetically contained plasmas, the densities are low, hence, substantial containment times of the order of seconds will be required. By starting with a solid-fuel D–T pellet, a very high n value is obtained, and only a very short τ, comparable with that inherent in the inertia of the interaction fuel following ignition, is required. Here again the reaction neutrons would be used to breed tritium fuel in a lithium blanket.

It is still too early to be able to estimate with any degree of certainty exactly which approach to economic fusion power holds the highest promise of early, or of ultimate, success. However, progress in recent years has been so promising that significantly increased activity and support seem imperative in view of the national and social benefits that success in this field will bring.

The first and most important step in achieving this degree of success was the expansion of basic knowledge of plasma physics. Without this knowledge, either of two catastrophes could have taken place: (a) there would have been inadequate clues to the most productive lines of research, with a consequent delay in progress; and (b) had fusion power proved scientifically impossible on this planet, resources of time, manpower, and money would have been committed to countless costly trial-and-error experiments, with no hope of success. There is an element of trial and error in every experiment, but it is fundamental knowledge—mathematical theory and physical understanding—that guides experiments toward what is called a rational approach. In this way, the multidimensional infinity of possible experimental attempts is reduced to a finite set of logically related steps.

The validity of the scientific method for the solution of man's practical problems is so well documented that the further example of fusion research might seem redundant. Nevertheless, throughout the current pattern of support for science, there is recurring overemphasis on immediate practical relevance. The enthusiastic and optimistic attempts of the early 1950's to forge ahead to the development of a fusion reactor, without consideration of the gaping holes in the knowledge of plasma physics (many then totally unrecognized), were doomed from the outset. A major fraction of the knowledge that has guided the work in controlled fusion has been fundamental in nature; that is, the original creative thought did not arise in response to a specific short-range goal but, rather, stemmed from a desire to understand the nature of things. This pattern has implications for the development of any national policy for science, either basic or applied.

Fluid Dynamics

Thousands of problems of the physics of fluids have been and are being solved for the benefit of man. An enumeration of even part of these would constitute a telephone directory of scientific achievement. Testifying to these achievements are such stupendously successful engineering results as airplanes, ships, rockets, oil pipelines, the functioning of the Weather Service, and the understanding of the motion of galactic gas clouds.

Of principal concern in this brief review are the unknown and the structure of a science that continually pushes back man's frontiers to this unknown. Flow of fluids separates into two broad classes, laminar and turbulent. By and large, laminar-flow patterns, a flow that is smooth and steady, are now understood.

But when flow stream lines become distorted, convoluted, and mixed in that wondrous random pattern called turbulence (for an example see Figure 4.67), the fundamental knowledge of fluids breaks down almost completely. In leaning over the fantail of a ship and watching the complicated ever-changing patterns of the wake in the ocean, the mind is teased by the almost total unpredictability. The need to develop an adequate description of turbulence motivates the largest fundamental effort in the physics of fluids. To a certain extent, the lack of predictability of turbulence can be predicted, but the problem of quantifying such a notion still remains. Some examples of the inadequate predictability of unpredictability follow.

The problem of the motion of stirred tea leaves is familiar from allusions to the tempest in the teapot, but the general convective patterns in a rotating fluid are still far from solution. Rotating fluids with a turbulent boundary layer are involved in many forms of pollution separators, and the short-circuiting of the separation work by turbulent-flow patterns is a significant and unresolved problem. The most important of these separators is, of course, the atmosphere of this rotating planet, which encompasses the extraordinarily complex phenomena of cyclonic storms, thunderstorms, tornadoes, and smog. Each of these components may be many orders of magnitude removed from full understanding, but in total they constitute an overwhelming impact on our environment. Understanding of the electrifica-

[185]

FIGURE 4.67   Air flowing past a cylinder rolls up into a regular succession of vortices. A cross section of these vortices is made visible by injecting a sheet of smoke in the center of a wind tunnel. (Photograph taken by Gary Koopmann at the Naval Research Laboratory, Washington, D.C.) [Source: *Yearbook of Science and Technology 1971*. Copyright © 1971, by McGraw-Hill Book Company, Inc. Used with permission of McGraw-Hill Book Company.]

tion process and subsequent precipitation in a thunderstorm still is lacking. Arguments focus on whether a tornado requires electrical energy input to release its awesome power; understanding is insufficient to reach an answer.

Two-dimensional turbulence describes many of the large-scale atmospheric phenomena on this planet as well as on other planets. Two-dimensional turbulence also describes much of the surface circulation on the sun and the stars. Yet little is known about how to quantify and predict even the behavior of such two-dimensional situations. An important phenomenon resulting from two-dimensional turbulence is the jet stream that meanders across the continent at high altitude and affects so dramatically many transcontinental airplane flights, quite apart from weather effects. Another familiar example is the motion of a fish; in many cases, the fish uses far less energy to move than would be predicted from normal turbulence theory. Perhaps some particular waggle, developed in a long evolutionary process, somehow sheds a turbulent boundary layer more rapidly than otherwise would be the case. It is also possible that the skin of some efficient swimmers, such as the porpoise, can be neurologically and physiologically active instead of passive, so that it can damp incipient eddies in the boundary layer before they develop.

These, and many similar problems in turbulent flow, stimulate the imagination of many scientists, but the creativity and effort of the theoretical hydrodynamicist is challenged most of all by the need to understand the fundamental aspect of all these problems—namely, the behavior of the turbulent fluid. An increase in the knowledge of such a fundamental aspect of fluids could have wide impact and applications.

The current lack of understanding of thunderstorms, tornadoes, jet streams, clear-air turbulence, and other atmospheric phenomena affects the everyday life of society. With fuller knowledge, prevention of tornadoes in some cases might be possible. Or scientists might be able to establish a continuously operating thunderstorm in the Los Angeles basin and similar areas during times of temperature inversion through the imaginative use of the waste heat from a very large power plant. One such large thunderstorm is enough to wash and purify the air of the Los Angeles basin every 12 hours.

The addition of traces of long-chain organic molecules to fluid streams is already reducing turbulent drag by as much as 80 percent in certain cases; with fuller understanding of the way that these additives work, further improvements might be feasible, with a reduction in costs of pumping fluids and of transportation through and on the surface of fluids.

Bistable fluid jet devices, which can assume either of two states and which may be switched by a relatively small signal, may be used either in logic circuits or as power relays (see Plate 4.XII). As compared with electronic or electrical devices, they are not nearly so fast, but they have advantages in cost and in freedom from damage owing to heat (as in proximity to jet engines) or to radiation fluxes (as in the interior of nuclear reactors).

Plasma Physics Other Than Fusion Research

There are clearly important applied goals in plasma physics other than fusion. New fundamental knowledge in plasma physics has been applied also to the problem of direct conversion of electrical energy from hydrodynamic flow [the magnetohydrodynamic (MHD) generator] (see Figure 4.68). Progress has been made in MHD generation, using fossil fuels, to the point at which it appears practical to consider power plants considerably more efficient than the current ones and emitting significantly less pollution in the atmosphere. Very large MHD stations are now under construction in the Soviet Union for inclusion in that nation's national electrical grids. In the United States, MHD has been largely an industrial research problem,

WALL ATTACHMENT SWITCH

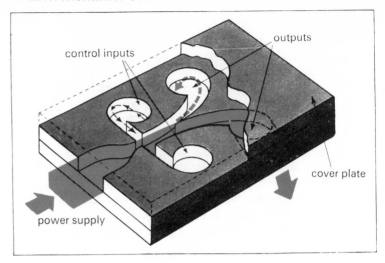

PROPORTIONAL AMPLIFIER

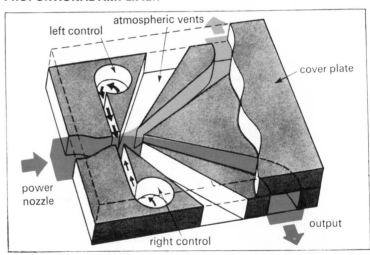

PLATE 4.XII Fluidic devices are frequently based on one of four basic concepts. The wall attachment switch will stabilize with the output wholly along one channel, depending on microscopic asymmetry in manufacture and other variables. But if the control nozzle on that side is turned on, fluid is entrained (black arrows) and the asymmetry at once switches the flow to the other output. Reblocking the control nozzle will not cause the flow to revert (broken gray line); it will stay attached to the new output. The turbulence amplifier has a supply tube and output tube in alignment. Flow from any control pipe at once converts the laminar main jet to turbulent flow, causing sudden and drastic reduction in pressure in

TURBULENCE AMPLIFIER

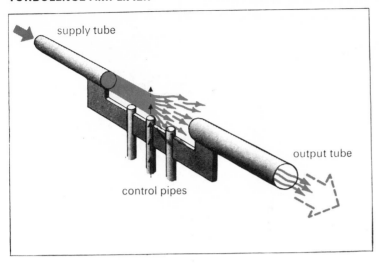

supply tube

output tube

control pipes

VORTEX AMPLIFIER

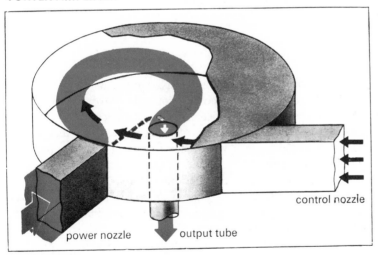

control nozzle

power nozzle   output tube

the output (original, full output shown as broken arrow). The beam deflection proportional amplifier divides its jet according to relative flows through left and right control nozzles. In condition shown, the left control is greater, giving higher pressure in the right output. The vortex amplifier offers little resistance when the control flow is switched off (main flow then following line of broken gray). With control pressure applied, the main jet is made to swirl round the vortex chamber so that resistance rises and flow is reduced. High control pressure reduces main flow almost to zero.

[189]

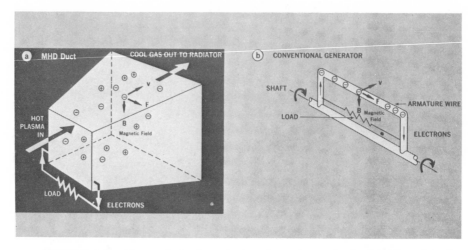

FIGURE 4.68   Magnetohydrodynamic (MHD) generator. In the MHD duct (a), the electrons in the hot plasma move to the right under the influence of force F in the magnetic field B. The electrons collected by the right-hand side of the duct are carried to the load. In a wire in the armature of a conventional generator (b) the electrons are forced to the right by the magnetic field. [Source: USAEC/Division of Technical Information.]

with federal funding sought on a matching basis. Unfortunately, the matching federal funds for the presently proposed pilot plants have not been made available, and the utility industry has been reluctant to invest in the development of a new power source in the face of the immediacy of the present power shortage. It would seem a wise federal policy to assure that new power sources, with a potential for dramatic improvement in efficiency and in the quality of the environment, are vigorously pursued.

The fundamental understanding of plasma physics will lead to a better knowledge of conditions beyond the earth—the magnetosphere, solar wind, solar corona, cosmic rays, and ionized clouds in galactic space. Most matter outside the earth's atmosphere is in the plasma state.

There is also a comparatively large research area of great potential impact—the study of the dynamics of highly ionized plasmas—in which the laboratory experimental work is beginning to expand in both the United States and the Soviet Union. It includes wave motions in plasma, turbulence (both MHD turbulence and electrostatic turbulence), shock waves, instabilities, the generation and emission of high-energy particles and nonthermal radiation, wave echo phenomena, and properties of relativistic plasmas. These phenomena, which are increasingly invoked in the inter-

[190]

pretation of astrophysical and space observations, are now being studied under controlled laboratory conditions.

The study of the dynamics of plasmas in magnetic fields has broad significance. Unlike most media studied by physicists in the laboratory, plasmas cannot yet be fully controlled and are highly sensitive to the exact details of the apparatus used to study and generate them, and especially to impurities. Although magnetic control of plasmas has been the main aim of thermonuclear research for many years, improvements in the understanding of plasma dynamics leading to better control of plasmas could have an impact in many other fields. For example, applying existing knowledge might lead to important developments in ion sources for ion propulsion and heavy-ion sources for nuclear research. Many suggestions have been made in the last ten years for accelerators incorporating plasma in one form or another; recently, groups in the Soviet Union announced the successful operation of the first electron ring accelerator (ERA) embodying some of these principles.

In the general area of low temperatures and partially ionized plasma, basic work is required that will improve the understanding of ionospheric physics and reinforce various technological applications of plasmas. Three areas are worthy of special attention. First, the discovery of gas lasers provides powerful sources of radiation over an enormous range of the electromagnetic spectrum, particularly in the three decades between 1-μm and 1-mm wavelength. Although the potential applications in this range are many, the understanding of the physical processes in many gas lasers is comparatively poor.

A further example is the interaction between plasmas and surfaces of condensed matter. Here technological devices have been developed and used for many years with comparatively little understanding of the basic processes. For example, there is much dispute and uncertainty about the basic physical principle of an arc spot, and spark erosion machines are used with little understanding (other, perhaps, than crude energetics) of how the metal is removed.

An essential feature of ionization-physics studies is the provision of information on elementary processes such as collision cross sections, reaction rates, and details of atomic and molecular interactions with charged particles and radiation. At low gas pressures, sophisticated studies with electron, atomic, and molecular colliding beams, which can be readily carried out in academic laboratories, will lead to new cross-section data that are urgently required in many technical applications of plasma physics.

The recent development of high-intensity coherent light sources will lead to advances in the understanding of the interaction of radiation with matter and of mechanisms of plasma production in gases and solids. Improved determinations of molecular structure will follow from investigations of the scattering of laser radiation in gases and liquids. Further high-power laser development and extensions from the red end of the spectrum toward shorter wavelengths in the ultraviolet are likely to make possible, with the aid of nonlinear optical media, intense sources of ultraviolet radiation that are urgently needed in other plasma studies.

Studies of the mechanism of the electric spark extended to high-gas pressures and large-electrode separations will provide information on electron–atomic ionization and electron–ion recombination. All these investigations will be associated with the development of sensitive electronic detectors, fast optical and electronic techniques, and computing methods. Techniques involved in the production, separation, and detection of fundamental charged particles produced by high-energy nuclear accelerators are also closely linked with developments in ionization physics.

The technological application of the results of basic studies in plasma physics are widespread. Applications include electrical communications, space-vehicle design, and the control and transmission of electrical power. To a large extent the technical long-term future of ionization and plasma physics will be concerned with the objectives of fusion power, plasma and ion propulsion, and the development of MHD power generation and direct conversion systems.

In the short term, improvements can be expected in a wide variety of devices employing gaseous electronic phenomena, ranging from optical frequency and microwave communication systems to ultrahigh vacuum pumps and gauges, arc rectifiers, and lighting equipment. The possibility of the full control of the basic physical processes that lead to an electrical discharge opens the prospect of suppressing discharges for times sufficient to effect electrical switching operations free from danger and unreliability. Studies of the microplasmas at separating electrical contacts can lead to improved switching technology in low-power applications. Associated studies of the mechanism of electrical breakdown in gases at high pressures could facilitate new developments in large circuit breakers and to improvements in high-voltage power cables using high-pressure gas insulation.

Applications in high-power technology are perhaps more obvious but no less important. Extension to yet higher power transmission voltages (greater than 1 MV) will require basic studies of the physical mechanism of the electrical breakdown of insulators of a wide range of media, including gaseous, liquid, and solid insulators, as well as of vacuum.

Use of Computers in Plasma and Fluid Physics

Plasma physics is the most complex of the classical sciences, and the mathematics needed to describe the phenomena are correspondingly

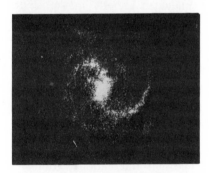

FIGURE 4.69 A computer's view of galactic evolution. Astronomers at the Goddard Institute for Space Studies, New York City, used a computer to simulate the growth of a galaxy. From top to bottom, the frames show the initial shapeless gas clouds, two growing spiral arms and two condensations, the start of a spiral pattern, and, finally, a well-developed spiral. [Source: *Science Year. The World Book Science Annual.* Copyright © 1971, Field Enterprises Educational Corporation.]

[193]

complex. For this reason, computers have played an especially important part in the development of deeper understanding of plasma and fluid dynamics phenomena. For an example, see Figure 4.69.

Since von Neumann's pioneering efforts in the use of computers in weather prediction, the problems of the fluid physics of the atmosphere and oceans along with nuclear weapons requirements, have excited the greatest pressure for the development of ever larger and faster computers. And the computer industry has responded. Experimental tests involve great costs, and in some cases they are physically impossible to carry out in complete fashion so that even partial replacement of them by numerical simulation requires large funding and high priority. This close coupling between plasma and fluid physics and computer technology will continue and even grow.

International Cooperation

Both plasma physics and the physics of fluids have benefited from, and added greatly to, the scientific knowledge of the world. They have been among the subfields of most vigorous international interchange and collaboration.

In the period from 1950 through 1958, controlled fusion research was conducted secretly in the United States, the United Kingdom, and the Soviet Union and to a much lesser extent in other countries, with major emphasis on the production of a fusion energy source. The increasingly evident long-range nature of the research task encouraged declassification in 1958. It was then found that very similar theoretical approaches and minor experimental achievements had characterized the different national programs. Soviet–U.S. relations at this time were personally amiable but distant and highly competitive.

The Geneva II Fusion Exhibits, particularly that of the United States, stimulated worldwide participation in controlled fusion research in 1958. The first International Atomic Energy Agency-sponsored conference on Plasma Physics and Controlled Nuclear Fusion Research was held in Salzburg and attended by 29 national delegations. The conference was marked by highly productive scientific exchanges, as well as by vigorous controversies along national lines.

Since 1961, the worldwide fusion research effort has moved toward real integration. There is a large-scale exchange of visitors and visitors-in-residence among participating nations, including the Soviet Union. Personal relations have become extremely cordial, and controversies, though still lively, form along scientific rather than national lines. The scientific value of this interaction has been so important for the U.S. fusion program that it is difficult to imagine what alternate course it might have followed in the absence of declassification. Certainly the cost of research to achieve comparable progress would have been much higher.

A similar impact on East–West relations can be claimed for meteorological research. During the original development of computer analysis of global circulation, the weather reporting of Soviet stations was crucial. The initial cooperation—a direct TWX line into the Lawrence Livermore Laboratory—has now been extended to the worldwide effort known as the Global Atmospheric Research Program, in which almost all the nations of the world are cooperating in an effort to understand weather on a worldwide basis. The impact of this and other international programs of cooperative meteorological research are of major importance to East–West relations.

Distribution of Activity

Work in plasma and fluid physics is widely distributed throughout the scientific and engineering communities. The major fraction of the PhD scientists working in these areas—6.7 percent of the physics PhD population as reported in the 1970 National Register of Scientific and Technical Personnel—is concentrated in the universities (48 percent), with substantial percentages found also in industrial laboratories (22 percent) and government and federally funded research laboratories (29 percent) (see Figure 4.70). A large part of plasma and fluids research is performed by people who consider themselves engineers or applied mathematicians. The combined physics research in plasma and fluids produces roughly 7 percent of the world's physics publications and 8 percent of the U.S. publications. Of the physics theses produced in 1965 and 1969, roughly 4 to 5 percent were produced in plasma physics and only 1 percent in the physics of fluids. On the other hand, 30 percent of those engineering theses (in *Dissertation Abstracts*) that might equally have been classified as physics were produced in fluid dynamics, so that the combined plasma and fluid theses constitute 17 percent of the total number dealing with the subject matter of physics, the contribution from fluids being about three times that of plasma physics.

The total annual funding of physics of fluids by the federal government amounts to approximately $37 million. By far the largest fraction of this support is motivated by military objectives; ships, planes, missiles, fluidic computers, and the like account for some $20 million a year. Much of this research is applied, but it also contributes to the store of fundamental knowledge. The DOD funds some work in universities, though much less than in previous years. The AEC, NASA, NSF, and NOAA each support the physics of fluids at a rate of about $4 million to $5 million a year.

The current funding for plasma physics is about $30 million. Nearly 85

PLASMA & FLUID PHYSICS

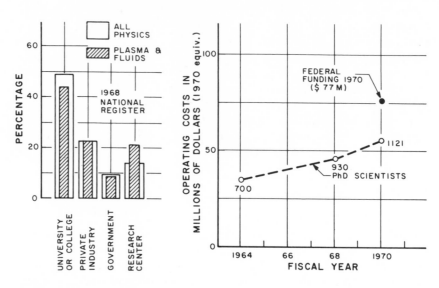

FIGURE 4.70   Manpower, funding, and employment data in plasma physics and the physics of fluids, 1964–1970.

percent of these funds are provided under the controlled thermonuclear research program of the AEC. The support of controlled fusion research in the United States is concentrated on large confinement experiments in a number of major laboratories. The universities have pursued smaller, but important, experiments, have trained young scientists for the major laboratories, and in turn have tended to recruit their professors from major laboratory personnel. Support of plasma physics as a basic science is largely the responsibility of the NSF. In 1970, approximately 17 percent of federal funds and 9 percent of industrial funds for basic research in physics were allocated to the support of basic research in fluids and plasmas (Figure 4.71).

Problems in the Subfield

In view of the great potential importance of controlled fusion research to the world, and of the steady progress being made toward the goal, the lack of reinforcement by increased support is surprising. By way of comparison, the ratio of U.S. to U.S.S.R. manpower in this area dropped from

PLASMAS AND FLUIDS

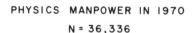

PHYSICS MANPOWER IN 1970

N = 36,336

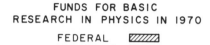

FUNDS FOR BASIC
RESEARCH IN PHYSICS IN 1970

FEDERAL

INDUSTRIAL

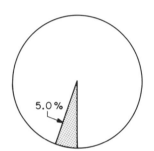

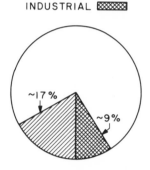

FIGURE 4.71   Manpower and funding data in plasma physics and the physics of fluids in 1970.

1 in 1960 to about 0.3 in 1970. Federal support over the past five years has been essentially constant, with a consequent significant reduction of manpower in the major fusion research efforts, particularly in the AEC laboratories. This reduction has the effect of reducing experimental capability at a time when the possibility of major advances has been demonstrated by other nations and when progress in the U.S. program has been marked. The nation cannot afford continuation of these downward trends, as one major recommendation in Chapter 2 for a stated national commitment in this area implies.

The current funding of plasma physics as a basic science is such as to reduce drastically the support of new proposals. The dollar investment by the NSF and the AEC declined in 1971; yet the motivation for federal support of research and new knowledge in this subfield is vastly greater now than before. The present funding of MHD is indeed dismal. Considering the very large application and relevance of plasma physics, there is need for supporting basic work at a much higher level than is presently the case.

The overwhelming relevance of fluids has so far maintained the level of funding in the physics of fluids. On the other hand, the near-constant funding for the last four years implies a fractional decrease of 20 to 25 percent in the net effort expended. Serious dislocations of the subfield will take place if present funding trends continue much longer. In view of the

[197]

fact that the foremost phenomenological aspect of fluids, namely turbulence, is not yet subject to a quantitative analytical description, such a reduction of effort seems indeed shortsighted. Compounding the gradual inflationary reduction of effort is the current political expediency of relevance. Application of this criterion to proposals dealing with a basic understanding of turbulence does a disservice to all concerned. First and foremost, it discourages work on the most important problem of fluids.

> We have next to speak of the stars, as they are
> called, of their composition, shape and movements.
> ARISTOTLE (384–322 B.C.)
> *On the Heavens,* Book II, 289

> Hence it is incumbent on the person who specializes
> in physics to discuss the infinite and to inquire
> whether there *is* such a thing or not, and if there
> is, *what* is it.
> ARISTOTLE (384–322 B.C.)
> *Physics,* Book II, 202

> When I, behold upon the night's starr'd face,
> Huge cloudy symbols of a high romance
> KEATS (1795–1821)
> *When I Have Fears*

ASTROPHYSICS AND RELATIVITY

Introduction

The interface between physics and astronomy is one of the most active in the physical sciences, and indeed the division between these fields is increasingly blurred and artificial. All of physics is required for any coherent understanding of some of the most recent and most spectacular discoveries in the heavens. Nuclear astrophysics has evolved into almost a separate area of specialization in nuclear physics, atomic physics has long played a central role in both observational astronomy and the study of stellar atmospheres and the interstellar medium, elementary-particle physics is now being called on to contribute to the understanding of the totally new high-energy phenomena at the hearts of stellar objects, and condensed-matter physics is needed to elucidate the structure of their cooler regions. The importance of plasma physics and optics in astrophysics is obvious.

Because a special relationship exists between astronomy and physics in regard to relativistic and gravitational phenomena and cosmology, and because developments in recent years have dramatically changed the nature of the physics in these areas from inspired and intuitive theoretical prediction to a more balanced situation in which these predictions are accessible to experimental test, astrophysics and relativity receive special consideration and discussion in this review.

One of the classic writings on astrophysics and relativity begins with the subtitle, *The Bigger They Are, The Harder They Fall*. It could well serve as a motto of this subfield, which deals with very large bodies such as stars and galaxies, the strong gravitational forces they create, and the way they behave under the influence of such forces. The mysterious force of gravitation was described accurately by Newton in the seventeenth century, and his classic work on the subject has been the basis for virtually all calculations about the way bodies move under the influence of gravitation, from the orbits of planets and, more recently, space probes, to the slow turning of the vast systems of stars, known as galaxies, in space.

But Einstein, in 1915, proved that gravitation can be reinterpreted as a curvature of space–time. The power of this concept lies in the finding, noted by Galileo and Newton, that in ordinary empty space bodies not acted on by forces travel in straight lines; in Einstein's space–time, curved by gravitation, bodies still travel the shortest path between two points, but this path is now curved. Technically, the curvature of space–time refers to the distortion of geometrical figures formed by light rays between various points. For example, in a curved space, the angles of triangles no longer sum to 180°, just as is the case for triangles drawn on the curved surface of the earth. But the concept of shortest distance— like the great circle route between cities on the earth—is still valid, and, according to Einstein, this is the type of path actually pursued by bodies under the gravitational forces exerted by a star, planet, or galaxy.

Although Einstein enunciated his theory early in this century, it has had little impact on the thinking of most physicists and astronomers, let alone nonscientists. The reason is not hard to find. Newton's conception that space itself is flat (described by normal Euclidean geometry), and that bodies in the solar system travel on curved paths within this space, yields highly accurate calculations. Indeed, the three classical tests of Einstein's theory of general relativity present such extremely small deviations from Newton's classical theory that they are still the subject of continued discussion and measurement.

A sequence of discoveries in astronomy, starting in the post-World War II era, using a variety of new techniques such as radio astronomy, have brought to light a variety of extraordinary phenomena—radio galaxies, quasars, cosmic background radiation, and pulsars. In each of these cases, in contrast to previous experience, scientists could be dealing with systems in which the differences between the theories of Einstein and Newton are profound. Relativistic astrophysics brings together astronomers, who need a theory such as Einstein's to interpret what they see, and physicists, who are eager to test the validity of Einstein's theory and to show how it applies in cases where space is very much curved.

Application to Cosmology

Because the effects of gravitation are important only for large amounts of material such as are found in a star (whose typical mass is one million times that of the planet earth), one might expect major effects to be observed on the largest possible scale in the universe. Indeed, Edwin Hubble in the 1920's formulated the remarkable law of the red shift, according to which galaxies seem to be receding from the earth at enormous speeds that are greater as the distance of the galaxy from earth increases. Friedmann showed that this phenomenon is a natural consequence of Einstein's theory; thus the science of relativistic cosmology—the application of relativity to the universe—came into being. Hubble's observations had indicated velocities up to about 20 percent of the velocity of light. Relativistic cosmology predicts that at even greater distances material would be seen to be moving away from earth at much higher speeds. Only recently, a quasar was found (a strange astronomical object described later in this review) that appears to be receding from earth with a speed nearly 90 percent that of light. Moreover, a variant of Friedmann's calculations, one frequently identified with Gamow and known as the big-bang model of the universe, predicts that radiation should be reaching earth from the very limits of the universe, where matter is receding from earth at a speed differing from that of light by only one part in a million. (In connection with this phenomenon, see Plate 4.XIII.) Recently, this prediction was partially confirmed by measurements at radio wavelengths. Radiation emitted as visible light from the limits of the universe arrives at the earth as radio waves (this is simply a consequence of the familiar Doppler effect).

Pulsars, Supernovae, and Black Holes

There are two ways to obtain a strong gravitational field. One is to have a large mass, which, of course, is the case with the universe. Another is to have a smaller mass, such as that of a star, and to come extremely close to it, which can happen only if the mass is compressed into a very small volume. Some stars have extraordinarily small diameters compared with those of normal stars such as the sun. White dwarfs, which are a rather common type of star but too feeble to be seen without a large telescope,

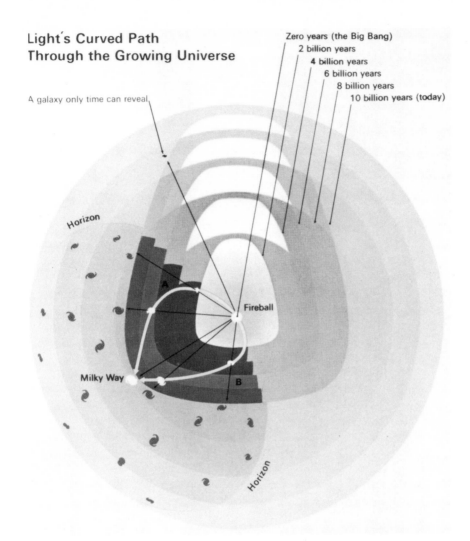

Light's Curved Path Through the Growing Universe

Zero years (the Big Bang)
2 billion years
4 billion years
6 billion years
8 billion years
10 billion years (today)

A galaxy only time can reveal

Horizon

A

Fireball

Milky Way

B

Horizon

PLATE 4.XIII Light's curved path through the growing universe. The universe has swelled since its genesis in the primordial fireball. The two-dimensional spherical shell is shown at 2,000,000,000-year jumps. The light from only relatively nearby galaxies has had time to move across the shell and reach us. In fact, the oldest light we can see, that of the fireball itself, has reached us from only as far as the horizon. As we look out to the horizon, then, we see galaxies at earlier times, when the universe was smaller and they were closer to us. Their light must have taken the same curved path as the runner on an expanding race track. The light from every one of the billions upon billions of galaxies we see today must have come to us along curved lines like A and B. [Source: *Science Year. The World Book Science Annual.* Copyright © 1967, Field Enterprises Educational Corporation.]

[201]

are only 1 percent of the size of these normal stars. Strong gravitational fields have been found for these white dwarfs; however, extremely strong gravitation was expected only in the case of what was a purely theoretical concept—the neutron star—now believed to be the cause of the phenomenon known as the pulsar. In 1967, pulsars were found quite by accident as radio sources in the sky whose intensity flickers rapidly but with extreme precision. The initial reaction to the observed precision was that it represented a signal from extraterrestial intelligences. The most famous of these sources, the pulsar in the Crab nebula, pulses every 1/30 sec with extreme regularity (see Figure 4.72). The basic phenomenon is apparently a rotating star that has one or more bright spots on it. The radiation from the star then appears to vary as the bright spots pass overhead. To fit this interpretation, the star must have a diameter of only 1/100,000 that of an ordinary star but a comparable mass. Under such conditions, as Oppenheimer and Volkov had shown 30 years before, ordinary matter is crushed into nuclear matter, with a density 10^{14} times that of water. Space–time in the neighborhood of a neutron star must be extremely curved if Einstein is right, and calculations are now under way to predict what phenomena might be observed as a consequence.

A neutron star could be formed as a result of a nuclear explosion occurring inside an ordinary star. The core of the star collapses to the fantastic density of nuclear matter, and the outer layers leave the scene at tremendous speeds, causing the phenomenon that astronomers have long known as a supernova. In fact, the Crab nebula, first observed by court astronomers of the Ming Dynasty in China on July 4, 1054, is probably the remains of the supernova explosion that left behind the pulsar that is spinning rapidly in the center of the nebula today. (The concluding section of this chapter describes the neutron star and its structure.)

There is evidence that the Crab nebula is filled with fast electrons traveling at nearly the speed of light. Probably they were accelerated to their enormous speeds by a whiplash effect of the spinning neutron star. Since such fast electrons are observed throughout the Milky Way galaxy, it is quite possible that they originated in other supernova events similar to those in the Crab nebula. This hypothesis offers a solution to the riddle of the cosmic rays—fast protons and electrons that have long been known to bombard the earth and were man's first source of very-high-energy particles.

If one takes seriously the predictions of general relativity, and the example of the red shift is convincing, an extraordinary dilemma arises in regard to the fate of certain types of star. Generally stars are suspended in a delicate balance between their radiation energy, which continually escapes to space, and the compensating energy generated deep inside them

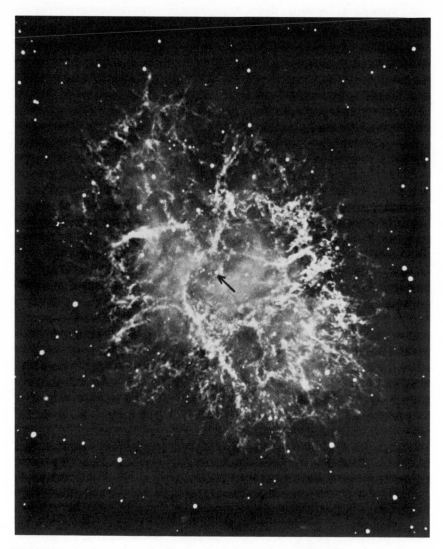

FIGURE 4.72   The Crab nebula. Arrow marks the pulsar NP 0531. (Lick Observatory photograph.) [Source: *Yearbook of Science and Technology 1971*. Copyright © 1971, by McGraw-Hill Book Company, Inc. Used with permission of McGraw-Hill Book Company.]

by thermonuclear reactions. Ultimately, a star must exhaust its nuclear fuel; when it does so, it attempts to compensate its radiation loss by releasing gravitational energy in a slow contraction. The dwindling radiation pressure from the interior no longer balances the gravitational forces, and the star contracts. In the case of moderately massive stars, this process may lead to the formation of a supernova and a pulsar. But very massive stars, when they reach the stage at which a neutron star might form, have such strong gravitational fields that ordinary matter is crushed completely. The star goes into what is called a gravitational collapse, which, at least theoretically, brings all the matter in the star to infinite density at a geometrical point of vanishing spatial dimensions. The gravitational field of such an object is so strong that light waves either originating inside it or attempting to get past it to distant observers are sucked in and disappear forever, with the result that the object appears to be a black hole. Although no black holes have yet been detected unambiguously, theorists fully expect them to exist. Some possible candidates have been located in certain binary systems. To discover them requires techniques capable of detecting an absolute black void a few miles across, located trillions of miles away in space—this truly is a challenge!

It should be emphasized that this prediction of infinite density faces a most important paradox that can have far-reaching consequences for all physics. When a physical theory predicts infinite parameters in this sense, it simply indicates that the theory is inadequate and that major new physical insight must be brought to bear on the problem.

Some 60 years ago, physicists faced a similar paradox in the atom. Accelerated electrons were known to radiate electromagnetic energy, but still atoms, with electrons in closed orbits—and therefore subject to acceleration—were stable. Resolution of this paradox required the discovery and development of quantum mechanics to replace the classical arithmetic; this discovery revolutionized physics and much of philosophy.

It may well be that a new discovery of comparable importance awaits physics in the phenomena of gravitational collapse.

It has recently been suggested that in this process of gravitational collapse a star that is not perfectly spherical will emit a special kind of wave predicted by Einstein—a gravitational wave. Such waves, invisible and extremely hard to detect, should have the property that, as they pass the earth, all material objects are simultaneously squeezed in one direction and stretched in the other. Weber and others seeking to detect such waves have set up apparatus capable of recording displacements as little as 10^{-14} in. It appears to be recording such waves about three times a day, but this discovery is so difficult to explain that confirmation is urgently needed. These waves are much more frequent than had been anticipated; if this finding is correct, it may mean that gravitational collapse to a black hole is a rather common occurrence in the universe.

The discovery of the pulsars and the convincing identification of a neutron star as their source have given great impetus to what previously had been inspired speculation concerning gravitational collapse phenomena. Nothing is yet known about the behavior of matter at the inconceivable densities that would be achieved in the latter stages of collapse nor about the validity of natural laws as the dimensions of this remarkable system shrink without limit. And what happens to the star after it collapses to a point? Physics currently has no answer. This situation challenges some of the ablest minds. It is truly at the brink of the unknown.

Galaxies and Quasars

Radio galaxies and quasars were first detected as a result of their radio emission; subsequent observations have shown them to be extremely bright emitters of visible light as well as of x rays and infrared radiation. The total amounts of energy radiated are stupendous—for example, in one case, at least 10^{11} times the radiation from the sun. At first these objects were related to normal galaxies, which are collections of individual stars. Normal galaxies emit up to 10^{11} times as much energy as the sun, but still only 1/1000 as much as a bright quasar. The extraordinary aspect is that quasars are much smaller than ordinary galaxies, and it seems impossible to explain their radiation in terms of many individual stars. Particles moving in a sufficiently strong gravitational field can emit a huge amount of energy. Up to 50 percent of the theoretical absolute maximum energy ($E=mc^2$) should be possible, but nuclear reactions in stars yield only 0.1 percent or less. Perhaps quasars represent the first known energy source powered by general relativistic collapse. But it has also been suggested that at least some of them correspond to collisions between matter and antimatter components of our universe, with the observed radiation originating through total annihilation processes.

Eventually the enormous energy production by quasars may be attributed to gravitational collapse; this is perhaps the key significance of the discovery of quasars for physics. However, there are awkward problems, even if it can be shown that the basic energy source is gravitation. Much of the electromagnetic emission in all wavelength regions, including radio, infrared, optical, and x-ray wavelengths, is believed to be synchrotron emission, a type of radiation that was first predicted and found in nuclear electron accelerators called synchrotrons. When an electron, a tiny electrical particle, moves at speeds near that of light in a magnetic field, so that it is accelerated into a circular orbit, it emits a peculiar glow, which, in the synchrotrons where this effect was first noted, is concentrated in the ultraviolet part of the spectrum. This emission of radiation by an accelerated charge is a classical phenomenon predictable on the basis of Maxwell's equations.

Because of various clues, astrophysicists are certain that the same emission process operates in quasars and radio galaxies (as well as in the Crab nebula). The radiation comes from a huge number of very rapidly moving, accelerated electrons. The question is, just how can these electrons be accelerated to the required speeds, even granting that a powerful energy source is available? In the laboratory, it is necessary to build an elaborate machine, such as a synchrotron, to accelerate electrons to the required energies. It is difficult to imagine how even a collapsing star, or possibly a collapsing galaxy, would produce great clouds of fast electrons. This problem falls within the scope of plasma physics, or its specialized offspring, plasma astrophysics—the subfields that study charged particles and their interactions. Solar flares—explosions on the surface of the sun—which are known to emit copious quantities of fast electrons, provide some clues. This phenomenon is known to be associated with magnetic fields in the solar atmosphere; since magnetic fields are also present in quasars, it could well be that these phenomena are related. But to show how the gravitational energy of collapse can be converted into the electromagnetic energy of synchrotron radiation remains a most challenging problem.

It is not only the general relativist and the plasma physicist who are interested in quasars but also the astronomer studying the evolution of galaxies. Quasars and radio galaxies may form from part of a class of objects in distant space * of which the normal galaxies are but another part. According to one interpretation of quasars, they are associated with the birth of a normal galaxy, the enormous explosive energy found in them coming from the massive stars that are expected to form and die explosively early in the life of a galaxy. A contrasting but equally arguable interpretation suggests that quasars are associated with the final throes of a dying galaxy, in which stars, drawn by the relentless force of gravitation to the center of their galaxy, are packed so closely that they begin to collide with one another, with disastrous results for them and a correspondingly high energy output. Whether the outburst of a quasar is the birth pang or the death rattle of a galaxy remains obscure, but astronomers and astrophysicists are trying to find out which, if either, is correct.

---

* If the red shift turns out to be gravitational rather than cosmological in origin, these objects are much closer to the earth than is now commonly believed.

Relation to Man's Fate

What have such exotic phenomena as quasars and pulsars to do with man's understanding of his own history and significance? The key idea in understanding this relationship is evolution. Anthropology discusses the cultural evolution of man from the time he first emerged from a primitive state. Biology discusses the evolution of plants and animals over a much greater time scale, culminating in man. Thus, natural selection, operating over millions, and even billions, of years, has led simple one-celled organisms to evolve into the extraordinarily complex phenomenon we know as man. Recent studies have shown that living matter was present on earth as long as 3 billion years ago. It is believed that the earth is only 4.7 billion years old; and since older life forms are constantly being found, it is possible that life may have originated on earth soon after its creation. The very recent discovery of materials that are essential for life—the amino acids—embedded deep inside a meteorite suggests that prebiological material of the sort that would be needed for early life on earth may already have been present in the solar system at the time the earth was formed.

It is believed that life formed in a natural and predictable way on the earth, and it seems likely that a similar phenomenon has occurred throughout our galaxy, in which there are billions of stars that are similar to the sun in light and heat and composed of the same chemical elements found on the earth, which are necessary for life. This being the case, life could well be ubiquitous throughout our galaxy and in many distant galaxies as well. Literally billions of life systems may exist independently throughout the universe. Some of them may have evolved to a state of intelligence, such that one can contemplate communication with them. Although no such communication has yet occurred, in the minds of at least some dreamers it seems a reasonable possibility. As Condon has discussed, however, unless we grant these intelligences the wisdom to prolong their civilization for periods much longer than 100,000 years—a wisdom that we seem as yet to lack—the probability of our actually achieving this communication appears vanishingly small simply because of the vastness of even our own galaxy. Intrinsic in any such statement is a trace of human arrogance; for it assumes that velocities and channels of communication that are now inconceivable to human physics do not, and cannot, exist.

The point of this discussion is that evolution toward higher forms of life appears to be an inevitable consequence of the uniformity of nature as we now understand it from astronomy. Not only are individual stars very similar to each other, with the chemical elements from which they are built in almost exactly the same proportion from one star to the next, but other whole galaxies seem to be very similar to the one in which we live. This situation would seem to call for a special explanation, until one realizes that the physical processes operating in the formation of stars or planets and of chemical elements appear to be the same throughout the universe. The extraordinary fact is that the atoms that can be detected by sensitive means, even in radio galaxies so far away that they are receding from earth at half the speed of light, appear to be identical to those found in a sample of dirt from a nearby lot. That this is far from coincidence is explained by the physicists who study the nuclei of atoms and the elementary particles from which they are made. They have shown in countless terrestrial experiments that all atoms on the earth belonging to a given chemical element are interchangeable, and that the same is true of the individual particles of which these elements are made. Furthermore, if Einstein was right, the gravitational forces that govern the evolution of stars, planets, and galaxies, and the collapse of interstellar clouds to form stars and planets from the dust floating in the interstellar space inside of galaxies, are also of uniform character throughout the universe. Relativistic cosmology, according to which material billions of light years from the earth moves under the same laws of gravitation as that on earth, rests entirely on this principle of uniformity. Possibly, astrophysicists will be able to surmise the entire evolutionary history of the sun, the earth, the biosphere (living matter on the surface of the earth), and man from observations of distant but similar systems that could be in earlier phases of evolution. This procedure is used in stellar astronomy, in which stars both younger and older than the sun, but otherwise having the same characteristics, are observed. In the case of galaxies, the study of quasars as a possible birth or death stage of galaxies may provide a clue to the fate of the Milky Way system. Consequently, astrophysicists continue to probe the past—although not in the time–travel fashion of science fiction—for clues to the future.

The Origin of Things

How did this whole evolutionary cycle begin? Currently there is no definitive answer; perhaps there never will be. A simple, yet ultimately obscure, theory apparently agrees with presently known facts in most respects. The big-bang or Friedmann model of the universe suggests that all matter from which galaxies and stars are made originated in a catastrophic explosion at the origin of time as we measure it. The density of matter was infinite at that original moment, but the matter was expanding outward so rapidly that its density quickly became low and normal galaxies could form. Two lines of reasoning, both associated with the fact that the explosion must have been very hot, provide evidence for this theory. At the high temperatures involved in the first second or so of the life of the universe, many reactions between atomic nuclei would necessarily have taken place and certain of the chemical elements would have been formed. Calculations indicate that the two main elements produced would have been hydrogen and helium, light gases familiar on the earth in their application to lighter-than-air flight. About 90 percent of the atoms would have been hydrogen, and about 10 percent would have been converted, according to this theory, into helium in the very early expansion phases. Study of the stars and interstellar matter confirms this hypothesis.

Another consequence of the very high temperature of the early universe would be cosmic blackbody radiation. Astrophysicists assume that this radiation originates at a surface far from the earth that was of high density and had not yet formed into galaxies at the time the light left it. In essence it is seen as it was just after the birth of the universe. This surface has a temperature about that of the surface of the sun, but, because the radiation it emits is red-shifted by the Doppler effect to 1000 times longer wavelengths, it can be perceived only by radio telescope and not by the human eye, as in the case of the sun. Again, observation generally confirms prediction in this respect. Einstein's theory of relativity is capable of describing the expansion of matter and the radiation in the early universe, once they had been created and sent on their way. However, it cannot explain the explosion or the initial presence of the matter or energy, and therein lies an extraordinarily frustrating puzzle.

The assumed uniformity of physical processes in the universe is based on the observation that the elementary particles involved are identical throughout. Apparently these elementary particles were created in the initial instant of the expansion; this event cannot be explained by elementary-particle physics at the present time. Relativistic cosmology and elementary-particle physics come together in the study of this initial catastrophe. An ultimate goal of relativistic astrophysics is the unraveling of the phenomena that led to the early explosion. (What happened before it is, at present, in the realm of metaphysics.)

A less ambitious goal is to understand the phenomena in the universe that occurred after the explosion, including formation of the chemical elements, formation of galaxies, and the relationship of quasars and radio sources to distant galaxies. There is good evidence that the elements heavier than hydrogen and helium, including the carbon, oxygen, and nitrogen of which living matter is composed, were formed inside stars. When certain stars such as supernovae explode, they eject these elements into interstellar space, where they are available to form new stars and planets. Thus, the study of supernovae and pulsars is linked to the origin of the elements. One can even go so far as to speculate that all mankind is composed of atoms that were once inside an exploding star such as that which gave rise to the Crab nebula.

In summary, the picture of cosmic evolution that fits much of the present data is this: Following an initial and inexplicable big bang in which the elementary particles were formed, the universe expanded and cooled until hydrogen and helium formed. Later these elements condensed into galaxies in which stars formed that were capable of further nuclear "cooking" to produce heavier elements. These elements, ejected into space by supernovae explosions (see Figure 4.73), formed more stars and planets, which are the habitats of widespread life forms throughout the universe. It is a

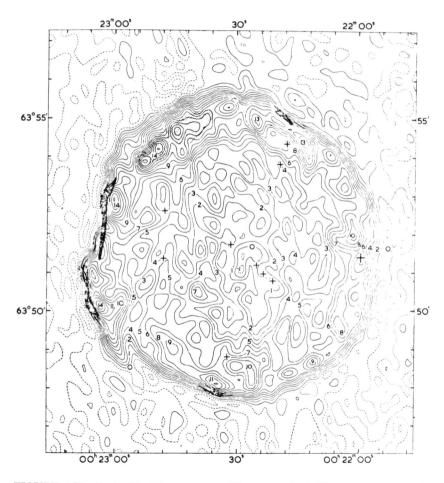

FIGURE 4.73 Tycho Brahe's supernova. The remnant of this supernova was observed by aperture synthesis at 1407 MHz with the 1-mile interferometer at the Mullard Radio Astronomy Observatory, Cambridge. [Source: J. E. Baldwin, "Radioastronomy and the Galactic System," IAO Symposium 313 (1967).]

major triumph of the human intellect that, on the basis of a knowledge of physics, the results of laboratory measurements, and mathematical skills, it is possible to reproduce, with remarkable precision and detail, the life cycle of the universe, assuming, of course, the initial explosion.

Tests of General Relativity

In addition to the question of the origin of the universe, astrophysics and relativity seek to work out the fundamental laws that govern the universe, even though many of them may not be of immediate application in the astronomical world as currently perceived. Thus, the theory of general relativity predicts that already in the solar system there should be minute deviations from Newtonian physics. So far no one has shown that these deviations are of any practical importance, but as a matter of principle it is important to search for the small deviations that are predicted to find out whether Einstein was right, for the correct gravitational theory has implications for the evolution of the universe. For many years attempts have been made to do this, but as yet with mixed success. Now new methods using radio and radar astronomy are being brought into play, with the hope that the precision of measurements can be made vastly greater than was previously possible and that a decision between the Newton and Einstein theories can be made once and for all. Present precision is already great enough to be confident that Newton was wrong in principle, but measurements still are not accurate enough to distinguish between two versions of the theory of general relativity—Einstein's and a variant, called the scalar–tensor theory, developed more recently by Dicke and Brans. It is hoped that further measurements in the solar system will give a final answer to this problem and thereby give physicists confidence in applying one or the other of these theories to distant matter in the universe.

Ways of Testing General Relativity

The number of scientists who are actively engaged in studying relativity *per se* is rather small, less than a few hundred. On the other hand, the equipment required for their studies is expensive, with the result that a substantial fraction, perhaps one fourth, of the budget devoted to the study of astronomy and astrophysics is allocated to research in astrophysics and relativity. For example, the United States has radio telescopes of varying sizes, ranging from 10 to 1000 ft in diameter (see Figure 4.74). The larger installations are expensive to operate and rely heavily on the federal government for support. As telescopes are multipurpose instruments, capa-

[212]

FIGURE 4.74 The radio–radar Arecibo Ionospheric Observatory in Puerto Rico. This giant has many uses, ranging from studies of the earth's ionosphere to the most distant radio sources. [Source: *Science Year. The World Book Science Annual.* Copyright © 1966, Field Enterprises Educational Corporation.]

ble of receiving radiation from any part of the sky and at a variety of wavelengths, they are used to study a wide variety of objects, ranging from the atmosphere of Venus to the radiation of distant quasars. However, because of the great interest in objects such as quasars and pulsars, a very significant fraction of their time, and consequently of the associated support budgets, may be considered as devoted to the study of relativistic objects. If one looks only at the expenditures of the scientists who are studying general relativity, one finds a total expenditure by the federal government of a few million dollars per year. On the other hand, if the various astronomical investigations that are related to this subfield are included, the expenditures may be ten times higher.

The nature of the enterprise involves theoretical physicists and astronomers, working with paper, pencil, and computer to delineate the implications of Einstein's theory; it also involves a much larger number of physics or astronomy PhD's who work with a variety of instruments to detect radiation from distant sources. Before World War II, the only such information available was from optical telescopes in the western part of the United States, many of them built with private funds, such as the 100-in. telescope on Mt. Wilson. Since the war, federal funds have been used to build a variety of optical telescopes in this country, with two new 150-in. optical telescopes scheduled for completion in 1973. Their power to penetrate space will rival that of the justly famous 200-in. telescope on Mount Palomar (see Figures 4.75 and 4.76.) In addition, smaller instruments have been built and used for the exploration of infrared emission by distant quasars. Because the atmosphere inhibits such observations, the telescopes are often mounted in airplanes or balloons. National observatories funded by the federal government account for a sizable part of the effort in this subfield; they provide larger instruments than any one institute or university can support. These instruments are then used by university personnel, as well as by the staff of the observatory, to study the faint signals from space on which the science depends.

In recent years, rockets and satellites above the earth's atmosphere have been used to study those wavelengths that cannot penetrate the earth's atmosphere, particularly x and gamma radiation. Such studies have shown

FIGURE 4.75   200-in. Hale telescope pointing to zenith; seen from the east. (Photograph courtesy of the Hale Observatories.)

FIGURE 4.76   NGC 5194, spiral nebula in Canes Venatici. Photograph taken with the 200-in. telescope. (Photograph courtesy of the Hale Observatories.)

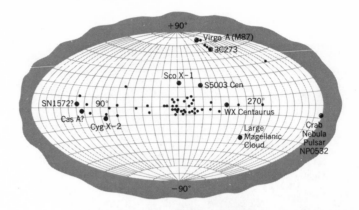

FIGURE 4.77 X-ray astronomy. *Top:* Distribution of the cosmic x-ray sources plotted in the galactic ($/^{11}$, b^{11}) coordinate system. Most of the sources lie close to the equator and are in the Milky Way galaxy. Two sources, Virgo A and the Large Magellanic Cloud, have been positively identified with other galaxies. Encircled sources have been optically identified.

Bottom: X-ray telescope to study solar x-ray phenomena from Skylab. The x rays enter the narrow annulus in the front and are focused by glancing-incidence fused-silica mirrors. (NASA photograph.)

[Source: *Yearbook of Science and Technology 1971.* Copyright © 1971, by McGraw-Hill Book Company, Inc. Used with permission of McGraw-Hill Book Company.]

that a nearby galaxy noted first for its radio emission emits even more energy in the form of x rays. All the instruments used are expensive. A large radio telescope costs approximately $10 million, and a large optical telescope costs a comparable amount. Space instrumentation can be even more expensive, because of the requirement of operating the facility by automated means at a great distance from the earth and because of the rocket power that is necessary to put such devices into orbit. The detailed instrumentation needs for relativistic astrophysics have been examined in the Report of the Panel on Astrophysics and Relativity (see Volume II) and are the subject of further remarks in Chapter 5 as well as many of the recommendations of the Astronomy Survey Committee of the National Academy of Sciences.

Present Status and Future Hopes

Progress in this subfield has been little short of astonishing, given the instruments that are available for detecting faint radiation. Optical astronomers have used the large West Coast telescopes to study the physics of quasars and galaxies. Infrared astronomers have made very significant measurements with their still rather modest instruments. X-ray astronomers, using rockets and orbiting observatories, have surveyed the sky and pinpointed interesting x-ray sources (see Figure 4.77). Radio astronomers have utilized radio telescopes in various parts of the United States to survey a large number of radio sources.

In the years ahead optical astronomers will bring into operation the two new 150-in. telescopes already mentioned, which will yield a new stream of information, particularly if this information can be processed at the focus of the telescope with the most recent electronic devices. Infrared astronomers also plan a significant increase in aperture capabilities. X-ray and gamma-ray astronomers are placing their hopes on a series of satellites, particularly a high-energy astronomical observatory, which will embrace a whole series of x-ray and gamma-ray experiments in one satellite. Since this satellite will operate for a year or more and survey the entire sky instead of being confined to a few minutes of observation, as are rockets, an enormous amount of new information on this subject should result. Radio astronomers for some time have planned a series of new instruments that will permit them to form a clearer picture of the distant radio sources that have been detected previously. So far it has not been possible to obtain funds for the largest of these, an array of radio telescopes capable of forming a fine beam for probing distant space, but it is hoped that this lack can be remedied in the next few years.

[217]

Distribution of Activity

Astrophysics and relativity, defined as the application of general relativity to astronomical systems, is actively pursued at comparatively few institutions in the United States and abroad. With a broader definition, one finds that much of contemporary astronomy and astrophysics, a significant amount of theoretical and experimental physics, and a considerable fraction of the space program are related in one way or another to this subfield. Several hundred physicists and astronomers are members of special divisions of the American Astronomical Society and the American Physical Society that deal with subjects in the scope of this subfield. Work is proceeding in dozens of astronomy and physics departments across the country, as well as at several research centers maintained by federal funds.

Einstein's theory is extraordinarily complicated mathematically. Because, until recently, there were few problems subject to observations for which it seemed necessary to use this theory, its study was primarily carried on by mathematicians, although a few theoretical and experimental physicists scattered about the country maintained an interest in it. But with the discovery of the extraordinary astronomical phenomena mentioned above, greater effort has gone into applying the theory to realistic models of stars and galaxies. As a result, relativity is no longer a mathematical plaything but a practical tool for finding out what is taking place in the universe.

Astronomers and physicists engaged in work in this subfield find it exciting, for it touches on fundamental questions of the evolution of matter and life. To obtain fundamental proof that some version of general relativity is correct would provide a new philosophical framework preferable to that developed by Newton in the sixteenth century, in which events in space–time, related by inexorable laws, give rise to the extraordinary structure and beauty of the astronomical universe. Increasing numbers of physicists engaged in other subfields, such as nuclear and plasma physics, are attracted to astrophysics and relativity by the opportunity to use their specialized knowledge in solving such enigmas as the pulsar. Substantial funding is required for such research. The support of the general public must be gained if this work is to progress.

Unlike most subfields of physics and astronomy, astrophysics and relativity is not widely recognized as a distinct discipline. The Panel attempted to identify a core group whose research is primarily in this subfield and to identify those federal funds that clearly are supporting the efforts of this group. It also attempted to estimate the additional funds allocated to the support of observational programs that are used less directly by this group, even though these funds have not been specified by the federal agencies as being for this purpose. The estimated percentage of federal funds for research in this subfield (Figure 4.78) includes about one fourth of the total ground-based astronomy program ($12 million per year). When space-based research is included, total support is about $60 million per year. This estimate is quite uncertain. The cost per PhD, of course, reflects the high cost of space experiments. Without the space component, which provides crucial observations of distant relativistic objects, the cost per PhD drops to a figure corresponding more closely to the level of support in most physics subfields.

Problems in the Subfield

Astrophysics is continuously rocked by a stream of astonishing discoveries —quasars, x-ray sources, cosmic radiation, pulsars, and infrared galaxies. Hints of bizarre phenomena such as black holes are on the horizon. These dynamic phenomena, interpreted in terms of Einstein's theory of curved space–time, are leading to a new conception of the universe.

Recent discoveries, following on the opening of the whole electromagnetic spectrum by new techniques, were made by pioneer telescopes, which can be enormously improved by a new generation of instruments now coming off the drawing boards.

ASTROPHYSICS AND RELATIVITY

PHYSICS MANPOWER IN 1970

N = 36,336

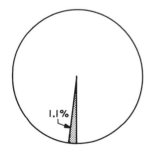

FUNDS FOR BASIC
RESEARCH IN PHYSICS IN 1970

FEDERAL

FIGURE 4.78   Manpower and funding in astrophysics and relativity in 1970.

[219]

The goal now is the construction of these major new facilities, including a large radio array, a variety of infrared telescopes, digitized imaging devices for major optical telescopes, and a high-energy astronomy observatory in earth orbit. The manpower to perfect and apply these instruments is already available.

That exciting discoveries about the universe will continue to be made cannot be doubted. Other nations have responded by committing major resources to this subfield. The United States must now decide whether it will grasp the opportunity it now has to participate in an all-out effort on the physics of the universe.

> They take place in the regions nearest to the motion of the stars.—It studies also all the affections which we may call common to air and water and kinds and parts of the earth.
> ARISTOTLE (384–322 B.C.)
> *Meteorology,* Book I, 338

EARTH AND PLANETARY PHYSICS

The Nature of the Subfield

Earth and planetary physics is a rather general title that encompasses a wide variety of scientific activities of ever-increasing importance in which physics and physicists play a significant role. The principal concern in this review is the use of the methods and concepts of physics in a number of research areas that deal with large, natural systems that generally cannot be controlled or altered by the observer.* Instead of applying the familiar techniques of laboratory investigation, the physicist studying such systems can only observe and record natural events (see Figures 4.79–4.81). His prime objectives are to understand the observed events and to predict their possible future occurrence.

Two other aspects of this subfield distinguish it from the rest of physics. First, money and effort are invested in studies of the earth and in certain aspects of planetary physics because they relate directly or indirectly to the relationship of man to his physical environment. The questions asked and the investigations pursued, although basic, have strongly applied overtones. For example, in regard to the earth, the prediction of hail, understanding of earthquake damage, analysis of storm surges, reliable

---

* However, many laboratory experiments are performed to study microscale phenomena, for example, water-droplet formation.

FIGURE 4.79   A submarine volcano erupts near Iceland. [Photograph courtesy of Photoreporters, Inc.]

FIGURE 4.80 *Top:* Barnard Glacier and Mt. Natazhat and Mt. Bear from the south southwest.

  Bottom: Mt. St. Elias looking northwest from over Malaspina Glacier.

  [Photographs courtesy of Bradford Washburn.]

radio communications, discovery of material resources, and better weather predictions are major objectives. There are, of course, both long- and short-term approaches to these problems. As an illustration, almost all research in weather modification thus far has been directed toward understanding the physical bases for weather phenomena rather than toward attempting weather-modification operations, which, in the absence of such understanding, could well have disastrous consequences.

There are many reasons for studying the physics of earth, and although there is a strong aesthetic appeal in understanding phenomena that surround us, the roots of the subfield and the ultimate justification for the high level of activity that it enjoys are securely tied to the practical needs of society. This is in contrast to the physics of planets (planetology), which is concerned with the interiors, surfaces, atmospherics, and satellites of planets and is motivated primarily by scientific curiosity and interest arising from opportunities in the nation's space program. The search for extra-terrestrial life belongs to this category, because so much depends on physical techniques and physical phenomena.

Another distinguishing aspect of the subfield is its interdisciplinary character. It is not a closed subfield of physics, although physicists can contribute to much of the activity—much more, indeed, than they have in the past. Its problem-oriented programs and its large-scale projects have brought together a loosely allied community of physicists, chemists, engineers, and technicians with a wide variety of specializations. Each group is dependent on the others and must learn the discipline of working with colleagues whose attitudes and backgrounds differ and with problems whose hallmark is complexity.

The subfield falls into several separable components. Meteorology and atmospheric physics have to do with the lower regions of the atmosphere, with an emphasis on the study of weather and climate. Aeronomy is concerned with the chemical processes and electrical effects in the upper atmosphere, out to the magnetopause where the earth's atmosphere merges with the solar wind.

Oceanography embraces a wide range of disciplines that include marine biology. The subdiscipline of physical oceanography is concerned with global ocean circulation, currents, waves, temperature, salinity, and other physical features.

Solid-earth geophysics embraces a variety of disciplines that relate closely to geology, mineralogy, petrology, and the like. Some of the main concerns here are such areas as: geomagnetism (study of the magnitude, fluctuations, and origins of the earth's magnetic field), seismology (study of the propagation of elastic waves in the earth and of earthquakes), tectonics

[223]

FIGURE 4.81   Deep ocean swell. [Photograph courtesy of Jan Hahn, Woods Hole, Massachusetts.]

(the secular motions of the earth's crust and of the underlying mantle), and the physical and chemical properties of natural minerals at high temperatures and pressures.

All branches of atmospheric and solid-earth geophysics are reflected in planetary studies. Although remote from this environment, planetary investigations provide a different look at problems of vital interest to man and, as in the case of the recent studies on the moon, can effectively give a view of the earth at both earlier and later phases of its history. Comparative studies of the earth and the planets complement each other and lead to new approaches and to fresh ideas about fundamental processes.

Finally, as a product of the space age, an entirely new field of study—that of the interplanetary plasma—has appeared. This is related directly to the established physics subfields of cosmic rays and plasma physics; therefore, departments of physics have shown more of an interest in it than in most of the other components of this subfield. The term space physics often is used to describe this subject, as well as the upper atmospheric (aeronomy) studies requiring rockets and satellites.

Historical Perspective

The history of the subfield starts in the distant past, with the ancient interest in terrestrial phenomena and the solar system, but its character has continually changed in response to the demands of society and technical development and opportunity.

Aristotle divided the physical sciences into physics and meteorology, the latter embracing all beneath the orbit of the moon and corresponding roughly to what we now call physics of earth. He understood that physics would embrace the more orderly branches of nature but that meteorology would have to deal with complicated phenomena in which descriptive methods, as opposed to the more abstract processes of induction and deduction, would play an essential role.

In the last two centuries, conventional physics and studies of the sea, atmosphere, and earth have drawn apart. New experimental methods led physics rapidly toward goals concerned with the fundamental nature of matter and radiation, while industrial needs coupled to new technological advances, such as the telegraph and deep-drilling rigs, led studies of the earth to increasingly technical and descriptive methods, with little emphasis on fundamental understanding. These trends are now beginning to reverse, and the proper role of physics in earth and planetary investigations is becoming clearer. Deep understanding of the natural phenomena has become recognized as a necessary step to prediction, understanding, and control, just as it has in other subfields such as plasma physics. One of the primary means of reaching this understanding is through physical research.

A few milestones in the recent history of the various disciplines that contribute to the study of earth and planetary physics are the following:

Meteorology began to move rapidly with the invention of the telegraph and accelerated with the demands of the aircraft industry during and after World War I. Theories of fronts and air masses were the mainstay of the

[225]

subject through World War II, although the technologies of collection and dissemination of meteorological data improved continuously. Since 1950, the outstanding advance, spurred by von Neumann's insight, was the introduction of the computer, coupled with an increased understanding of the physics of atmospheric processes. Now meteorologists are beginning to exploit a combination of satellite (see Plate 4.XIV) and computer technology and are perhaps on the threshold of modifying climate and weather.

Aeronomy began with the discovery and study of the ionosphere, by Appleton and by Breit and Tuve, in the 1920's. Advance was slow until rockets became available after World War II. Only in recent years, however, was the systematic use of satellites for aeronomy research proposed. The separation between meteorology and aeronomy now appears artificial, and the distinction may disappear in the future.

Physical oceanography is a relatively young science. Studies of ocean tides are not new, but most other aspects of the subject are related to modern dynamical meteorology. In many schools the two subjects are taught together, and research workers cross readily between fields. This subject, too, appears to be on the threshold of rapid advance as more oceanographic research ships become available, new instruments are developed, and satellite technology is used.

Seismological stations have existed for a long time, but solid-earth geophysics did not become a major area of study until the years between the two World Wars when the needs of the petroleum and mining industries gave it impetus. This increased activity was paralleled by the gradual evolution of the belief that the present and past state of the earth could be understood in physical terms. The work of Bridgman, Birch, and others on the physics and behavior of matter at high pressures was a cornerstone in this respect. Since World War II, the origin and behavior of the earth's magnetic field on a geological time scale have emerged as major research topics. The Vela program for detection of underground explosions gave major impetus to seismology, and the gradual increase of observing facilities on ships together with more sophisticated instrumentation and methods for coring the deep-ocean sediments led, in the past few years, to the concept of sea-floor spreading. There is now a completely new picture of the earth's crust, in terms of large mobile plates floating on the underlying mantle, and through their impinging motions producing continental drift, sea-floor spreading, mountain building, and zones of seismic and volcanic activity. Even more important, experiments here and abroad suggest that with better instrumentation of geological faults and a better understanding of earthquake mechanisms it may be possible to predict the occurrence of earthquakes. Russian scientists believe that they can make ten-day predictions of earthquake magnitudes and locations in the Tadzhikistan Province. There is also some hope for preventing earthquakes (when their foci are shallow) by injecting fluid to relieve the strain along rock fractures.

Physics of the interplanetary medium is a direct product of space research. Its beginning can be placed a little earlier, but advance was not rapid until Van Allen's discovery of the radiation belts in one of the original International Geophysical Year (IGY) satellite flights in 1958.

A similar statement can be made about planetary studies generally. They have had a long history as part of astronomy, but advances were slow, and most astronomers professed little interest in earthlike objects. With the Mariner and Venera explorations of Mars and Venus and the Surveyor and Apollo missions to the moon, the subject changed rapidly and is now very actively studied, especially planetary surfaces, interiors, and atmospheres and the whole problem of the evolution of the solar system.

How Research Is Conducted

In its approach to observations, earth and planetary physics differs importantly from conventional physics; this difference is a crucial part of the character of the subfield. First, there is the problem of field observations. Since control of the atmosphere, the oceans, or the earth is not possible, it is necessary to measure them in three dimensions and follow the changes with time. Many problems arise. The observations can be difficult, even dangerous, and almost invariably they involve the discomfort and inconvenience of grappling with the environment. But this situation has its converse, for many scientists relish this closer contact with their environment.

Observations on a large scale are invariably expensive. Few research workers in this subfield would complain that the resources needed for research have been inadequate. However, the need to seek and justify large funds and to handle them responsibly are immensely time-consuming occupations, often involving political aspects. As in many other subfields that require large facilities and support, an effective scientist must be prepared to devote a significant fraction of his time to these essentially administrative matters.

The physical phenomena of the earth know no national boundaries, and, since the earliest times, investigations in this subfield have been international. A network of scientific organizations (principally under the International Union of Geodesy and Geophysics and the Committee on Space Research) and intergovernmental organizations (for example, the World Meteorological Organization) exists to initiate and coordinate research programs, to disseminate information, to aid less-developed countries scientifically, and to perform many other essential functions (see Plate 4.XV). Research programs have occasionally been such remarkable international successes that they have attracted the explicit attention and support of both diplomats and politicians. The IGY and the continuing cooperation in the Antarctic are two of the most striking examples of such international ventures.

Organizational and logistical problems associated with earth and planetary research, and the obvious interest of the military in some of its aspects, led to large in-house federal research efforts. For example, meteorological research is principally, but not exclusively, carried out by the Weather Bureau, now a part of the National Oceanic and Atmospheric Administration (NOAA) of the Department of Commerce. Partly to enable the universities to participate in big science, a number of national activities have developed: national research laboratories, research facilities, national programs, and data centers for dissemination of information. National programs are frequently organized by government agencies, sometimes also with international collaboration. The Arctic Ice Deformation Joint Experiment is an example of a largely national program involving universities, government agencies, and industry.

The National Center for Atmospheric Research (NCAR) is an example of a laboratory operated by a consortium of universities using government funds. The huge radio telescope at Arecibo, Puerto Rico,* is an example of a research facility, open to all competent researchers, operated by a single university—Cornell. The Greenbank radio-telescope installation is operated by Associated Universities Incorporated (AUI), the same organization that operates the elementary-particle, nuclear, and other facilities and programs of the Brookhaven National Laboratory.

Many of the observational data in this subfield are obtained for operational purposes. Meteorological information will be collected from ground stations, balloons, and aircraft for purely practical reasons regardless of any research requirements. The same is true for much ionospheric, seismographic, and oceanographic data. The existence of these data in reasonably processed form is one of the most important requirements for research in the subfield, and additional data centers are needed for their dissemination.

Similarly, studies pertaining to earth, studies of other planets, and satellites require the use of major astronomical observatories with optical, radar, and radio instrumentation, as well as major projects and programs such as Mariner, Pioneer, TOPS, the Grand Tour, and Stratoscope. Many of these studies require international collaboration, but usually at a lesser scale than the terrestrial research programs discussed above.

* Now the National Astronomy and Ionospheric Center (NAIC).

PLATE 4.XIV   Earth as seen from space. [Source: National Aeronautics and Space Administration.]

Weather Observation: Present and Planned

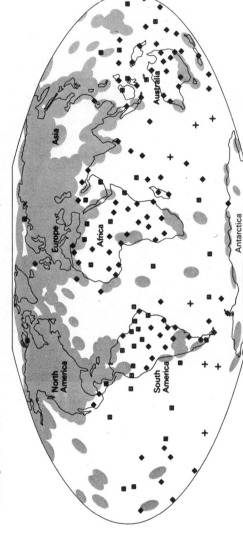

PLATE 4.XV Weather observation—present and planned. Deficiencies in weather observation stations are most noticeable in the Southern Hemisphere. The World Meteorological Organization has called for more equipment and increased observations at existing stations and for more surface and upper-air observatories. [Source: *Science Year. The World Book Science Annual.* Copyright © 1967, Field En-terprises Educational Corporation.]

Recent Achievements

Much more than conventional physics, this subfield is characterized by a gradual evolution of new concepts reflecting the combined thinking and contributions of a number of scientists. Discoveries of totally new phenomena are rare, as is the sudden emergence of new ideas. But it is also rare to look back over a decade without finding impressive advances.

In the past 20 years in physical oceanography, for example, major advances in both observing methods and theory have occurred. From the observational point of view the development of new sensors, or remote observing systems attached to buoys, and the use of numbers of ships to collect synoptic data have laid the foundation for a three-dimensional picture of temperatures, chemical composition, and currents in the sea. However, surprises still occur; after centuries of ocean travel and observation, not until the 1950's and early 1960's was the Equatorial Countercurrent discovered and described. This current is a remarkable narrow and shallow flow, with a maximum intensity almost exactly on the equator, moving in a direction *opposite* to that of the easterly trade winds.

On the theoretical side, methods developed by dynamical meteorologists have proved remarkably successful when applied to oceanic phenomena. There are now rather elaborate theories of the Gulf Stream, the Kuroshio, and the equatorial undercurrents and their relationship to the slower circulations in the main part of the oceans.

In the geophysics of the solid earth, the outstanding discovery of recent years is sea-floor spreading and its spectacular confirmation of the idea that continents are drifting apart, having once been joined (see Figure 4.82). The crust of the earth is now regarded as in a continual state of motion, with huge plates driven by upwelling currents from the interior.

Also of major importance has been the gradual evolution of ideas about the generation of the earth's magnetic field. That the field must be related to a dynamo motion in the earth's conducting, liquid interior is now known, although the precise mechanism has yet to be fully understood. Remarkably precise evidence shows that the direction of the earth's magnetic field has reversed many times, suddenly and for periods of varying duration, throughout geologic history (see Figure 4.83). Indeed, the record of these reversals written in the solidifying magma from midocean upwelling, can be read with as much confidence as tree rings in establishing a much older and more far-reaching chronology of the earth. Physical techniques have played a crucial role here in evolving magnetometers of ever-increasing sensitivity and convenience.

Meteorology since World War II has changed in almost all of its major

120 million years ago

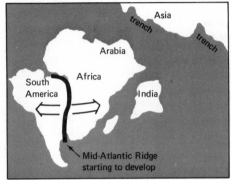

60 million years ago

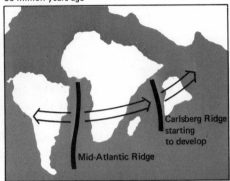

30 million years ago

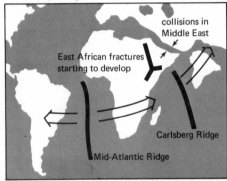

Present

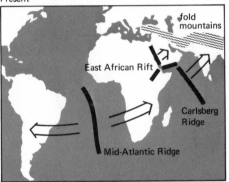

FIGURE 4.82 African and American continents probably started their drift apart some 120 million years ago. The initial separation of the combined African and Indian land masses from South America seems to have been caused by ocean-floor material spreading out of the mid-Atlantic Ridge. By about some 60 million years ago, the Carlsberg Ridge had begun to develop between Africa and India and helped to drive India toward Asia; one dramatic consequence was the formation of the Himalayas. About 30 million years ago, another ridge seems to have formed in the Gulf of Aden and, separated from the Carlsberg Ridge by the Owen Fracture Zone of transform faults, this has continued to develop independently. [Source: *Science Year. The World Book Science Annual.* Copyright © 1968, Field Enterprises Educational Corporation.]

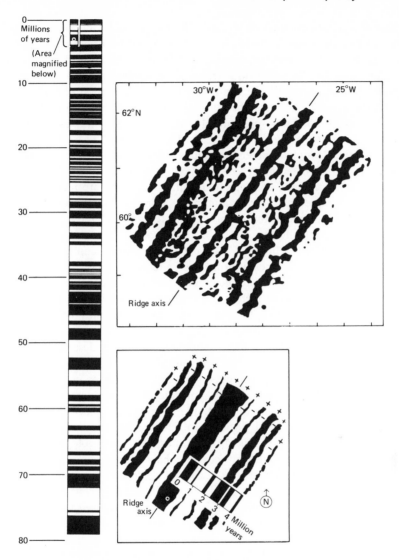

FIGURE 4.83   History of undersea magnetic reversals. The geomagnetic time scale (left) of 80 million years of magnetic reversals along one side of the mid-Atlantic Ridge matches an identical pattern on the opposite side of the Ridge. This time scale represents belts of rock of alternating polarity such as revealed in a map of Reykjanes Ridge, near Iceland (below). Each belt reflects the direction of the earth's magnetism while molten rock welled up through the Ridge rift. The magnetic map (top) has been clarified in the diagram (bottom). Note the close comparison between ridge reversals on the map and the geomagnetic time scale. [Source: *Science Year. The World Book Science Annual.* Copyright © 1968, Field Enterprises Educational Corporation.]

aspects. The mapping of upper air currents (notably the midlatitude jet stream) has given rise to studies of the stability of atmospheric motions and the upgradient transport of momentum. The outline of a theory of the general circulation is now available. Weather forecasting has been revolutionized by the advent of the computer, but the realization of this tool required the solution of many theoretical problems since the introduction of the idea by Richardson—prematurely, soon after World War I—and by von Neumann, following World War II. The last 20–30 years have brought about a rather complete understanding of the processes of precipitation, so that some forms of weather control now appear feasible. Fundamental questions of uncertainty of prediction in fluid problems are receiving attention. Theoretical climatology is imminent. Air chemistry is an established science and has provided the scientific basis for much of the recent concern about atmospheric pollution.

New laser-based techniques that permit extremely rapid and precise vertical density and composition profiles in the atmosphere, like those developed in oceanography, are beginning to make the construction of a realistic three-dimensional picture of the earth's atmosphere a reality. Much more extensive implementation of these new techniques is essential to further progress.

In aeronomy, a rather complete phenomenological description of the ionosphere has been developed. Chemical processes are studied in increasing complexity, so that the subject is now ready to come to grips with problems of upper-atmosphere pollution—questions of ever-increasing urgency as the density of supersonic high-altitude aircraft traffic increases. As dynamic transport processes by winds are included in the calculations, the subject slowly becomes more like lower-atmosphere meteorology.

Almost all recent planetary discoveries have come from space probes. Lunar samples have extended the history of the solar system back in time by a billion years. The surface of Mars (see Figure 4.84), its meteorology, and its aeronomy have become familiar. We are now poised for an attempt to detect life forms on that planet. Knowledge of Venus, with its surface at dull red heat, is evolving more slowly because of the ubiquitous cloud cover, but the Soviet and U.S. probes have reported back from this inhospitable surface (see Figure 4.85). The outer planets are less well understood because of their strangeness and remoteness, but here, too, ideas of composition and structure are developing at an increased pace.

Finally, the interplanetary plasma or the solar wind is mapped in detail near the earth, revealing a complicated system of shock waves and a hydromagnetic tail. The interaction of the solar wind with the moon and with Mars and Venus is also partially understood, and scientists now look toward the limits of the heliosphere, where the planetary system blends with the galactic system.

[234]

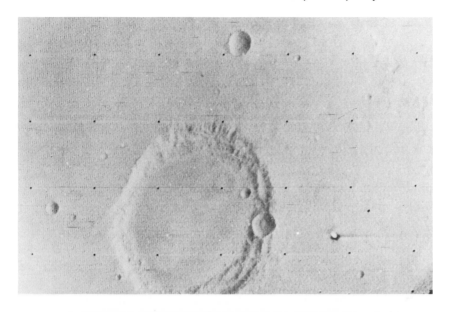

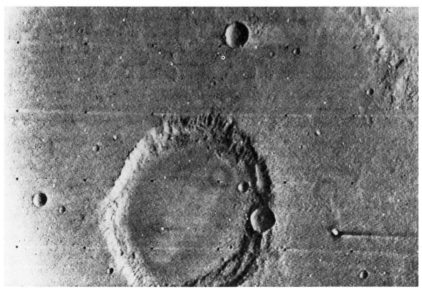

FIGURE 4.84 An illustration of computer enhancement of a Mariner probe photograph of the Martian surface. This is photograph 6N18 taken on July 31, 1969. The upper photograph is as received from the probe, and the lower is that obtained after computer enhancement. Enhancement is accomplished using the techniques developed at the California Institute of Technology Jet Propulsion Laboratory under NASA sponsorship and described by T. C. Rindfleish *et al.* in the *Journal of Geophysical Research, 76,* 394 (1971).

Spacecraft Measurements of Venus' Atmosphere

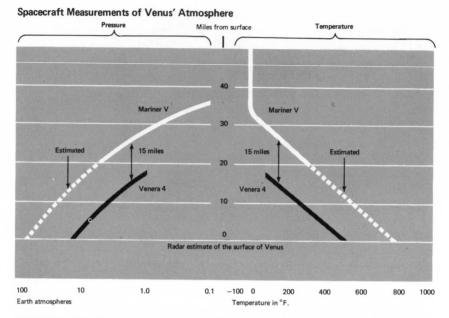

FIGURE 4.85   Russia's Venera 4 and the U.S. Mariner V returned near-identical data on the temperature and pressure of the atmosphere of Venus. The unexplained 15-mile gap could be from errors in earth-based radar measurements of the surface of the planet. [Source: *Science Year. The World Book Science Annual.* Copyright © 1968, Field Enterprises Educational Corporation.]

All of these developments have depended heavily on techniques and devices of physics, and the phenomena involved are those of classical physics, modern theory, condensed matter, hydrodynamics, optics, and electromagnetic theory, as well as those of the more modern subfields such as plasma physics and magnetohydrodynamics.

The Future of Earth and Planetary Physics

It is difficult to see limits to the development of this subfield. The pace of advance accelerates year by year, while the range of questions to be answered increases as applications become ever more apparent.

In physical oceanography, a better understanding of the transport of heat, mass, and momentum on both large and small scales should emerge. The large-scale transport couples with transport by the atmosphere and may be related to climactic changes. Transport by small-scale motions is important for the dispersal of pollutants. But the frontier of knowledge lies in the deep abyssal circulations. These have to be recorded and explained; advances in this area can be expected as the technology for abyssal study improves.

Meteorology should advance rapidly in the next decade, with the completion of the Global Atmospheric Research Program (GARP). The ten-day numerical forecast could well become a reality. And, importantly for many underdeveloped parts of the world, tropical meteorology should begin to catch up with midlatitude studies. The first major steps in weather modification could come in the next decade. Theoretical climatology and an explanation of climatic change on many time scales, including the ice ages, may be just over the horizon.

Solid-earth studies will change greatly as the ideas of plate tectonics and sea-floor spreading are exploited. The beginnings of earthquake forecasting have emerged. In this period an acceptable theory of the earth's magnetic field could be developed, as well as a clearer understanding of the seismological tool for studying the earth's interior. As increasing attention is paid to laboratory study of the properties of matter at high temperatures and pressures, studies of planetary interiors and of the early events in the solar system should become less speculative.

In space physics, measurements of solar and stellar particles in the solar ecliptic and beyond the limits of the solar system are anticipated. The purpose of the atmosphere Explorer missions will be to establish quantitatively the effects of specific processes that produce the F- and upper E-regions of the earth's ionosphere. Planetary exploration is likely to proceed at a rapid pace. Mercury will be reached, and its surface photographed. Probes and orbiters, tailored to a relatively modest budget, are likely to be launched to Venus. Mars will be the subject of extensive exploration, including the elaborate Viking lander that seeks to detect a new form of life. The first steps toward the exploration of the outer planets are now under way. Lunar studies may benefit from new unmanned techniques, although the Apollo program, after a remarkable record of achievements, is drawing to an end. All of these investigations are exploratory, and each of them should bring surprises and completely new insights.

Relationship to Societal Needs

Despite many new fundamental developments, the bases for support of this subfield rest, and will continue to rest, on the extent to which it can answer questions of more direct concern to mankind. Thus, in oceanography the main thrusts probably will relate to the production of food from the sea, the interaction of sea and coastline, the global distribution of water pollutants, and exploration and exploitation of the resources of the ocean floor. But the whole subject is in an early stage of development, and relatively undirected exploration of the motions and physical properties of the ocean at all depths must also be a feature of the studies conducted in coming decades. New insights emerging from fundamental studies will continue to generate opportunities for more applied investigations, but applications alone are neither a sufficient nor an efficient guide to the choice of problems for study.

In meteorology, the key problems will be, as they always have been, the prediction of weather and the possible control of weather and climate. Again, these questions can be answered only through deep exploration and understanding of fundamental questions. Problems of air pollution, which have long concerned meteorologists, will receive increasing study as their importance is brought to the attention of the general public.

In solid-earth geophysics, the traditional application has been petroleum and mineral exploration. With the rapid increase in the rate of utilization of these resources, locating additional reserves—often in less concentrated and accessible forms—assumes new and growing importance. Here the geophysicist, with his new instruments—magnetometer, earth conductivity probes, portable ^{252}Cf neutron probes, radiation-measuring equipment, and the like—plays a major role. To some extent emphasis will shift from local exploration to a more global geochemical census of mineral provinces. In addition to the vitally important goal of mineral exploration (scientific prospecting), the prediction of at least some features of the occurrence of earthquakes is a possibility.

Planetary exploration and space physics do not have direct applications to man's needs, except in certain specific cases. Radio transmission is still an important topic and was the original reason for the development of upper-atmospheric physics. There are obvious military applications of space research, and satellite techniques will continue to yield powerful new techniques for many kinds of terrestrial observation and monitoring, including resources. Further, planetary studies support, enlarge, and illuminate the findings of terrestrial studies. Although this is an indirect reason for support of planetary research, it remains a substantial consideration.

Manpower

A discussion of the role of physicists in the earth and space sciences presents a number of aspects that are not encountered in the traditional core subfields of physics. There are two major sources of difficulty. First, earth and planetary physics is not a separate, clear-cut field. Physics is one of the disciplines involved in the earth and space sciences. To be effective, the physicist working in these areas must combine his efforts with those of engineers, chemists, life scientists, meteorologists, geologists, and others. In doing so, his view of himself and the research process can change significantly from that engendered by his early training and experience as a physicist. Possibly, he will cease to identify himself as a physicist.

The second difficulty relates to the inhomogeneity of the earth sciences. Although all have some similarity in approach, the differences are real and important. Thus the geological sciences employ by far the largest number of scientists and differ from other earth and space sciences in having a large industrially employed segment and a large proportion of teachers. The geological sciences constitute a mature field, transfers to and from which are slow. Meteorology also is a mature field with low turnover of personnel, but industrial involvement is small and that in government-supported activities, high. In meteorology, research and development tend to be conducted by non-PhD's. On the other hand, both aeronomy and interplanetary research have always been populated largely by physicists. Data based on earth and space scientists in the societies composing the American Institute of Physics (AIP) tend to be weighted toward this group. Oceanography only recently has become an important field for physicists. It has a large interface with the life sciences. Both oceanographic and space sciences employ small numbers of scientists compared with the atmospheric sciences, which, in turn, are dwarfed in comparison with the geological sciences.

The main concern here is the relationship between the discipline of physics on the one hand and the earth and space sciences on the other. The interface defines the earth and planetary subfield. The relationship to physics usually is established by the nature of the highest academic degree, AIP membership, and self-identification. All three methods of identification are fallible; in particular, all can fail to identify a person whose research methods are those of physics but whose graduate department was not physics. Thus estimates of the number of physicists working in the interface are only approximate.

The 1968 edition of the NSF publication, *American Science Manpower,* provides data on those portions of the National Register of Scientific and Technical Personnel identifying the members of the American Meteorological Society and the American Geological Union. These data, together with 1968 and 1970 data on physicists identified with earth and planetary physics in the National Register survey, afford a partial description of the characteristics of scientists working in the subfield.

American Science Manpower 1968 shows that approximately 10 percent of the 298,000 participants in the 1968 National Register survey were identified with atmospheric, geological, and oceanographic disciplines, and that about 11 percent were identified with physics. About 1600 of the physicists work in the three former disciplines; however, very few of those in the earth and space science groups work in physics. A higher percentage of PhD's, composed largely of physicists, is found in the space sciences than in atmospheric, geological, and oceanographic disciplines taken together. In the space sciences, the percentage of PhD's (49 percent) is, as might be anticipated, close to that for physics (44 percent). Only 16 percent of the 23,160 scientists identified with atmospheric, earth, and ocean sciences in the 1968 Register survey held PhD's.

Scientists who indicated earth and planetary physics as their subfield of employment in 1968 numbered 1193, or 3.6 percent of the 32,491 physics respondents to the Register. In 1970, 1448—4 percent—of the 36,336 physics respondents to the Register were identified with earth and planetary physics. Exactly half of the physicists in this subfield held PhD degrees and comprised 4.4 percent of all PhD's in physics. The 724 nondoctorates in this subfield accounted for 3.7 percent of the total number of doctorates in physics in 1970.

The largest fraction of the PhD physicists working in earth and planetary physics is concentrated in the universities (38 percent); industry and government and federally funded research laboratories also account for substantial percentages (23 percent and 35 percent, respectively).

A high proportion of PhD's and non-PhD's in earth and planetary physics is engaged in research (67 percent and 64 percent, respectively). The comparable figures for all physics are 55 percent for PhD's and 47 percent for non-PhD's. In the earth sciences as a whole, the figures are 32 percent (PhD) and 18 percent (non-PhD). Thus patterns of activity and employment in earth and planetary physics differ from those that characterize earth sciences and physics.

Funding Patterns

In recognition of the importance of earth and planetary physics to societal needs, funding from federal sources has been relatively generous. Approximately $150 million per year is spent on research and research support, not including the NASA program. This sum divides in roughly equal proportions among meteorology, oceanography, and solid-earth geophysics; the major part of the expenditure is for work in government laboratories. Associated with this sum spent on research are far greater expenditures on operations and applications. The NASA program varies greatly from year to year, and the costs of hardware and development are difficult to evaluate.

These expenditures have generally been on a sufficient scale to support major demands, at least outside the space program. Promising new lines of research have usually been funded when the need arises. However, it is characteristic of most problems related to societal needs that the end is never in sight. Each improvement in weather forecasting leads to further demands: Man is rarely satisfied with existing capabilities until the point is reached at which the cost of advance is greater than the probable return. This last problem could be closer than scientists care to admit and should receive constant scrutiny.

The rate of advance is closely related to the level of support. The level of funding of research has not significantly changed in the last five years. However, the real value of this money has decreased. For example, progress in weather and climate control during this interval has been relatively slow. It is true that this is a very difficult area that contains complicated legal and political elements. However, the slow rate of advance should be a matter of greater concern than it has been.

In addition, solid-earth geophysics is entering a period of discovery and potential progress. Understanding of the dynamics of the earth's crust and the nature of earthquakes is growing. So far, federal agencies have not responded adequately to this situation. Perhaps the time has been too short, but adequate support for research is essential in the near future.

The final major question in funding patterns concerns the NASA program, particularly the planetary program. There is no area of knowledge that has been and will be more radically changed by the space program than that of the planetary system. The transition from the remote methods of astronomy to direct contact has created completely new fields of study for which the space probe is the unique tool of investigation. These fields are concerned with problems of historical importance to mankind that also have an impact on human needs.

Problems in the Subfield

The major problems in earth and planetary physics differ from one branch to another of this complex subfield. In most areas of the earth sciences the critically important task is to upgrade the research personnel so that they can undertake the extremely difficult physical problems that are now being revealed. The key to this problem lies in the universities. It is essential that earth and planetary physics be integrated in the physics teaching structure. By this means some of the excellent students entering physics programs will be able to appreciate the opportunities for research outside the laboratory and outside the core subfields of physics. The Report of the Panel on Earth and Planetary Physics (Volume II) offers recommendations on this and other subjects related to the educational problem and makes suggestions to foster the employment of more physicists in the earth sciences.

In the space sciences, the problems are fiscal and organizational. Again the Panel Report offers a variety of recommendations about the level of funding in space physics, the balance between large and small missions, NASA's advisory structure, the rationale for a space-science program, and other matters related to these questions.

> Chemistry is undergoing a renaissance. This is a new science. It is, perhaps, oldest of the laboratory sciences but it had fallen into the doldrums about the time of World War II; even the chemists found chemistry rather dull. Just after World War II the situation wasn't much improved. What has changed chemistry again has been the availability of a new set of tools. They have names which you may have read or heard of in one place or another: Nuclear magnetic resonance, spectro-photometry, very careful spectroscopy, electron diffraction, electron paramagnetic resonance spectrometry, instruments for measuring circular dictroism, mass spectrometry, and a few others combine to give the chemists a completely new set of handles on that aspect of the world for which they are responsible. This has permitted the chemists to understand the structure of molecules in a way no chemist understood them before. Instead of vague, two-dimensional chicken tracks written on paper, chemists now have a very clear understanding of the three-dimensional structure and electronic configuration of a large number of molecules as well as of chemical reaction on mechanisms.
>
> > Statement by Philip Handler, testimony before Daddario Committee, 1970 NSF Authorization Hearings, p. 10, Vol. 1 of transcript.

PHYSICS IN CHEMISTRY

Introduction

Throughout the long history of the physical sciences, physics and chemistry have maintained a very active and fruitful interchange of ideas and approaches. Progress has been complementary in the sense that major advances in one have opened new horizons or spurred new developments in the other. Interaction at the chemistry–physics interface is increasing rapidly. This discussion describes developments at this interface, which traditionally has been defined as *chemical physics* but includes other closely related subjects.

[242]

It is neither simple nor necessary to formulate precise definitions to distinguish physics from chemistry. Even so, it is useful to recognize certain characteristics that differentiate most of physics from most of chemistry. These are best expressed in terms of goals and attitudes, rather than explicit content. Physicists aspire to establish the *general* laws of behavior common to all of nature, while chemists establish the laws that distinguish one substance from another. Physicists tend to be more concerned with the common features of phenomena of matter; chemists, with those features that differentiate one species of matter from another.

Other distinctions between these disciplines have been suggested. Some people would, for example, restrict the definition of chemistry to studies concerned with extranuclear matter, so that nuclear structure would fall outside chemistry but chemical shifts in Mössbauer spectra would lie within chemistry. Much of what has been called the physics of condensed matter falls into the intermediate area between physics and chemistry because of its concern with materials and its attention to the distinctions among these substances. Yet even here, for example, condensed-matter physics tends to be concerned with the generic properties of crystals, while chemistry is more concerned with their classification and differentiation. Quite obviously, the interface moves with time; it is markedly different from what it was a few decades ago.

Subjects usually considered part of chemical physics include theory pertaining to wave properties of atoms and molecules; spectroscopy of atoms and molecules, from the radio-frequency range to some aspects of gamma-ray spectroscopy; chemical kinetics and collision processes, including hot-atom chemistry, radiation damage, and atomic and electronic collision processes at energies up to kilovolts or, perhaps, tens of kilovolts; the entire field of the structure of liquids (see Figure 4.86); polymeric molecules; statistical mechanics of both equilibrium and transport processes; molecular crystals, but only certain highly selected properties of other types of crystals, such as metals or semiconductors; constitutive thermodynamic properties of all phases of matter except perhaps superfluids, superconductors, and plasmas; and the use of lasers to study properties of matter (for example, nonlinearities). In the United States, these topics are frequently included within the scope of chemical physics; however, several of these subjects are considered pure physics in Europe and Japan.

Among the subjects that are included in the interface but are frequently excluded from chemical physics are properties of ionic, metallic, and covalent crystals and other periodic structure; amorphous materials such as glasses; semiconductors; the origin of the elements and several other aspects of nuclear physics, including many aspects of nuclear spectroscopy of radioactive species and some aspects of nuclear reactions; and superconductivity,

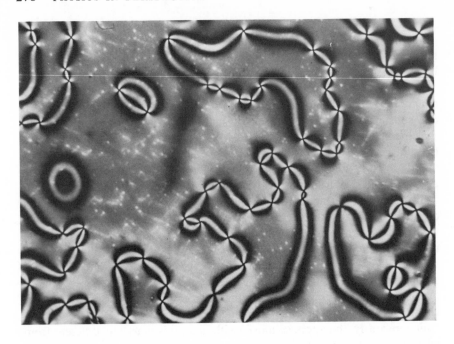

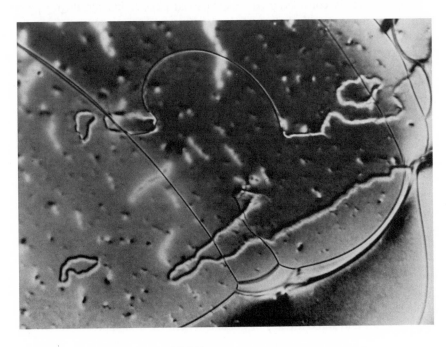

◄ FIGURE 4.86 Liquid crystals. *Top:* All the molecules along any black curve are aligned in the same direction in this polarized light view of the nematic crystal MBBA.

Bottom: This nematic liquid crystal seen in polarized light reveals the internal threadlike structures that are the result of spontaneous orientation of the molecules along their long axes. In this relatively thick specimen (about 100 μm) of MBBA, the thicker threads are on the surface, and the thinner ones are connected to the surface only at their ends.

[Photographs courtesy of Glenn Brown, Kent State University and the National Science Foundation.]

particularly with regard to its microscopic level of interpretation and the development of superconducting materials having specific properties.

Just as some subjects clearly relate chemistry and physics and develop quickly because of this relationship, some appear to suffer unduly slow development because of inadequate communication between these fields. The development of materials having specifically desired properties is one such area—for example, crystalline and amorphous solids having particular optical or electrical properties, liquids with particular viscous properties, or gas mixtures with special capabilities for energy transfer. As later examples will indicate, the communication between physicists and chemists dealing with properties of materials appears to be much more frequent and effective in industrial laboratories than it is in most universities, where too often there is relatively little communication between the two sciences.

The Influence of Physics and Chemistry on One Another

In some respects the relationships between physics and chemistry are asymmetrical. One illustration is the flow of manpower from subjects traditionally in one discipline to subjects associated with the other. A second illustration relates to the influence of one discipline on the other.

Although people trained as physicists in some instances practice what they call chemistry, movement from one discipline to the other generally is from chemistry, especially physical chemistry, into chemical physics or physics. Often these physical chemists and chemical physicists move into subjects in physics that pose challenging and important problems but currently receive little attention from the physics community. Possibly, the tendency for persons trained in chemistry to work on problems formerly regarded as belonging to traditional physics reflects the changing nature of chemistry and the delay between the time an approach is first used to solve chemical problems and the time when it is well assimilated into the chemical literature.

In regard to the influence of the content of one discipline on the other,

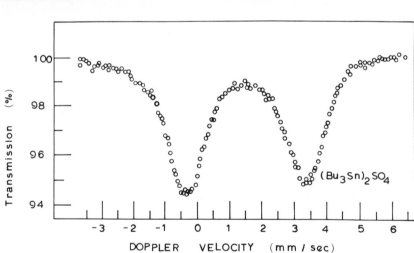

FIGURE 4.87　*Top:* Source–absorber–detector arrangement in a typical Mössbauer experiment. The radioactive source is attached to the velocity transducer on the right, which provides the energy modulation necessary in Mössbauer spectroscopy. The gamma radiation passes through the experimental sample, which is held at cryogenic temperature in the Dewar whose tail section is located in the center of the op-

we are clearly in a period in which most of the basic theoretical framework for chemistry and most of the chemist's physical methods for experimental research trace their origins to physics. Quantum mechanics, statistical mechanics, and thermodynamics—the entire theory of the extranuclear structure of matter—are essentially physical laws that underlie any theoretical interpretation of chemistry. Many (but not all) chemists now feel that the theory of the chemical bond has been reasonably well explained in terms of quantum mechanics. Fundamental and productive insights continue to occur in chemistry as a result of the application of quantum-mechanical methods to chemical problems. One particularly powerful tool, recently developed, is a set of interpretive and qualitative rules governing the spatial arrangements preferred by reacting species. These rules often permit the synthetic chemist to select from several possibilities a specific path by which he can control in detail the geometric arrangement of atoms within a molecule. He can, for example, find methods for synthesizing only the active form of a chemotherapeutic drug, instead of having the greater part of his product produced in inactive forms. Formerly, only natural biosynthetic processes exhibited this kind of efficiency. Only recently has it been recognized that the human olfactory system has a remarkable sensitivity to molecular structure such that the stereoisomers of the same molecule, except in rare circumstances, have strikingly different smells. There may be many more situations wherein the ability to control molecular architecture in detail can have important and practical consequences.

Experimental chemistry has been deeply influenced by physical methods. Spectroscopy, from the radio-frequency region of nuclear magnetic resonance through microwave, infrared, visible, ultraviolet, x-ray, and even some aspects of gamma-ray spectroscopy (see Figure 4.87), is vital to the chemist. Nuclear magnetic resonance (see Figure 4.88) has become an exceedingly powerful tool for analyzing chemical structures, because in most cases the response of a nucleus to this probe is highly characteristic of its immediate environment. For example, one can count the numbers of

---

tical path and impinges on the radiation detector on the left. A multichannel analyzer (not shown) stores the counting rate versus source–absorber velocity data for each cycle of the velocity transducer.

Bottom: Typical Mössbauer spectrum of the ^{119}Sn resonance (23.8 keV) in an organometallic compound. The presence of two resonance maxima (at ~ -0.3 mm/sec and at $\sim +3.3$ mm/sec) is due to the fact that the tin atom occupies a noncubic symmetry lattice site in the experimental compound. These data were recorded with the absorber held at liquid nitrogen temperature (78 K) in the Dewar tail section shown in the above figure.

[Courtesy of R. H. Herber, Rutgers University.]

[247]

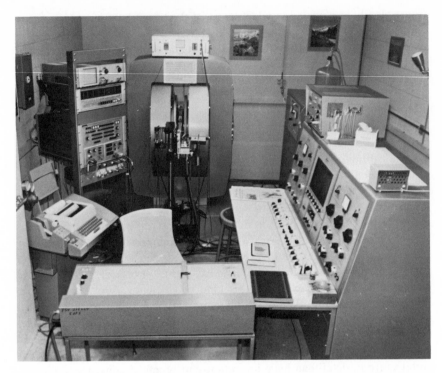

FIGURE 4.88   Bruker HFX-90 Multinuclear cw and pulsed FT NMR spectrometer consisting of (left to right) a Nicolet 1083 computer, magnet, and transmitter–receiver console. Variable-temperature and double-resonance capabilities are also present. [Photograph courtesy of the Department of Chemistry, Florida State University.]

hydrogen atoms in each of many kinds of site, even in rather complex molecules. The method is being developed and used now for the study of the structures of protein molecules. Electron spin resonance, in the microwave region, permits scientists to probe the location, hence the reactive site [as well as the structure (see Plate 4.XVI)] in the large class of highly active intermediates known as free radicals, the species responsible for the formation of many polymers. Microwave and infrared spectroscopy have enabled the chemist to study the sizes and shapes of molecules by letting him see how they rotate and vibrate. They also make possible the recognition of characteristic reactive parts of molecules for purposes of identification and monitoring. Ultraviolet and x-ray spectroscopy are tools for studying the electrons within the molecule; they permit the chemist to study, for example, the chemical bonds and the colors of substances. Chemists have

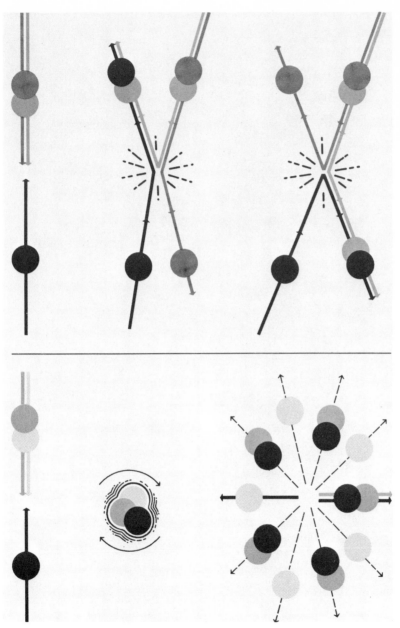

PLATE 4.XVI Determining complex intermediates from direct reactions. *Top left:* Colliding reactions may form a short-lived complex intermediate that rotates and pulsates with internal energy. *Bottom left:* The initial orientation lost, the complex decays into products that emerge in random directions. *Top right:* In a direct reaction between approaching molecules, one reactant may strip off atoms from the other in passing. *Bottom right:* Or it may collide head-on with the other and rebound attached to the atoms. In both cases, the directions of the emerging products are not random. [Source: *Science Year. The World Book Science Annual.* Copyright © 1968, Field Enterprises Educational Corporation.]

applied lasers in much of their work, including photochemistry, the study of energy conversion, and the probing of the energy levels of bound electrons.

Other techniques from physics have equally broad use. Mass spectrometry and x-ray crystallography are now necessary tools for the chemist.

One of the most striking recent transfers from physics to chemistry pertains to atomic and molecular beams. In this burgeoning chemical field the elementary atomic and molecular interactions, the electronic transfers and rearrangements involved, and the structural and energy dependence of the interactions are all subjected to detailed study. These studies for the first time reduce chemical kinetics to its elementary interactions. The entire arsenal of techniques, both experimental and theoretical, that have been developed in atomic and nuclear physics are thus being assimilated rapidly by chemistry. Most of the major chemistry departments now have chemical accelerators roughly parallel to the nuclear accelerators found in physics departments in the 1950's (see Figure 4.89). Through the use of crossed beams it becomes possible to extend the study of the chemical collisions to molecular species in highly reactive or metastable states. With merging beams it becomes possible to study these collisions at extremely low effective energies, as the relative velocity of the two reaction beam species is reduced toward zero. Adoption of this technique marks a major transformation of experimental chemistry from a macroscopic to a microscopic science.

Complementary to this activity is that of hot-atom chemistry. Here the time sequences of chemical reactions can be studied by following radioactively labeled atoms of carbon, nitrogen, fluorine, phosphorus, and the like as they move from one to another molecular site during the reaction process. Typically these labeled atoms, because of their characteristically short lifetime, are produced and studied in the environs of a large nuclear accelerator.

Many aspects of nuclear chemistry, as in the Lawrence Berkeley Laboratory, are not distinguishable from similar activities called nuclear physics. (In this survey all the nuclear chemistry activity *per se* is included in the discussions of nuclear physics.) Some 15 percent ($12 million) of the total current operating funds in nuclear science is designated for the support of nuclear chemistry.

It is extremely important to recognize a characteristic pattern associated with the assimiliation of any physical method into chemistry. When the method is first discovered, a few chemists, most often physical chemists, become aware of chemical applications of the method, construct their own homemade devices, or work with a physicist in an informal fashion to

FIGURE 4.89   Apparatus for crossed-molecular-beam experiments examining the reactions of alkyl and aryl halides with alkali metal ions at low kinetic energies. [Source: Photograph courtesy of the Department of Chemistry, Florida State University.]

demonstrate the utility of the new tool. At some point, according to the pattern, commercial models of the device are put onto the market. These are sometimes superior, sometimes inferior, to the homemade machines in terms of their ultimate capabilities to provide information. However, the commercial instruments generally are easier to use and far more reliable than the homemade devices. The impact of the commercial instruments is rapidly felt and often very far reaching; it sometimes virtually revolutionizes a field. Chemists with little interest in learning elaborate new techniques from physics can now apply a physical tool to answer questions of direct interest to them without an unreasonable expenditure of time and learning effort.

The chemical accelerators noted above, with energies measured in tens to hundreds of electron volts rather than MeV, are excellent examples. So also are the nuclear magnetic resonance and laser spectroscopy systems.

Chemists with the new instruments typically have not been concerned with the principles or the development of the device; they have concentrated their efforts on extracting the chemical information that the device can provide. In optical, infrared, and radio-frequency spectroscopy; mass spectroscopy; and x-ray crystallography this pattern has been followed. In the past 20 years, the chemistry curriculum has added the subjects of nuclear resonance; electron resonance; rotational, vibrational, and electronic spectroscopy; and x-ray crystallography. Any reasonably trained chemistry student has learned enough of each of these techniques to be able to recognize their capabilities and to apply them to chemical problems for which they are appropriate.

Other experimental methods that have great impact on chemistry include radioactive tracer and activation-analysis techniques, which are employed in both qualitative and quantitative analysis. These methods use the properties of the nucleus in a way analogous to that in which infrared and ultraviolet spectroscopy use the properties of electrons and atoms. Some of the most dramatic applications of these methods are in the interfaces between chemistry and other disciplines. For example, radioactive dating methods (initially developed by Libby, a chemist) have become an integral part of archaeology. Neutron activation and x-ray fluorescence are also used in archaeology, as well as in the study of fine arts and in criminology. The first analyses of the composition of the moon were accomplished by a simple backscattering version of Rutherford's initial experiment in which alpha particles emitted by a radioactive source are scattered from a target material, in this case the lunar surface (see Figure 4.90). This experiment, flown on NASA's Surveyor series, was developed by Turkevich, also a chemist.

Naturally enough, when a physical method has proven its ability to give accurate and reliable chemical data, it usually finds a place as a monitoring probe, often in industrial process control. The oil industry could hardly exist as we know it now without real-time mass analysis of the contents of its cracking towers by means of mass spectrometers.

The flow of new physical methods for the chemist continues. Electron spectroscopy—the analysis of electrons knocked from the molecules of a sample, either by light, electrons, or other means of excitation—is a very new and powerful tool for studying the way electrons hold a molecule together. In this area, the first commercial instruments are just appearing, and one can watch the rapid growth of a new kind of chemical information. That electron spectroscopy is a very powerful probe and monitor for environmental pollutants already has been demonstrated; it combines extremely high sensitivity with reliability and relatively low cost (see Figure 4.91).

[253]

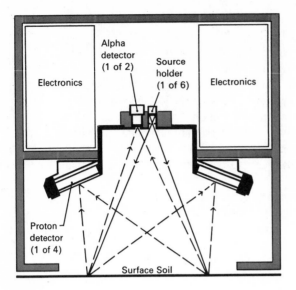

FIGURE 4.90 This alpha scattering instrument was designed to bombard a section of the lunar surface with alpha particles from radioactive curium-242. The energies of the backscattered particles provided information on the chemical makeup of the lunar surface. [Source: *Science Year. The World Book Science Annual.* Copyright © 1968, Field Enterprises Educational Corporation.]

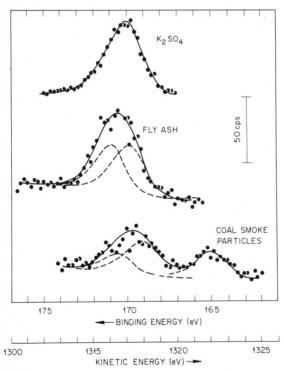

Photoelectron Spectra of Sulfur on Various Substrates.

FIGURE 4.91 Photoelectron spectrum of sulfur in fly ash and coalsmoke particles. The sulfate spectrum is included for comparison. These data are indicative of the presence of sulfur in various chemical forms, the higher energy peak being due to the presence of sulfate on the surface of the particle. [Photograph courtesy of T. Carlson, Oak Ridge National Laboratory.]

Another current example is the use of low-energy electron diffraction for the study of surfaces. This technique has important potential for increased understanding of the role of surfaces in catalysis; even small improvements here can be reflected in enormous industrial economies.

A third example is slightly less developed, but its promise is great; this is the use of very-long-wave infrared spectroscopy, actually interferometry, to study surface properties such as epitaxial growth. Chemists have used this method less than have physicists studying properties of solid materials; it illustrates the delay that can result from weak interaction between fields. Still another example is the application of Mössbauer spectroscopy to the study of dynamical properties of ions dissolved in water and other solvents.

Theoretical concepts, as well as hardware, continue to flow from physics to chemistry. Modern methods for dealing with interactions among electrons—problems of the electronic structure of atoms and molecules and of scattering of electrons by atoms and molecules (as well as molecule–molecule interaction)—have strong ties to the many-body methods developed in the context of nuclear and condensed-matter physics. In some cases, very closely related methods were developed independently by scientists, some of whom identified themselves as chemists and others as physicists.

Only recently have modern computer techniques been brought to bear on the enormously complex calculational problems involved in the microscopic quantum-mechanical descriptions of chemical reactions. Nonetheless, these calculations have already evolved to a point at which they are among the most demanding in all of science in terms of computer speed, size, and sheer number-crunching ability. Clementi and his associates at the Large Scale Computation Laboratory of IBM have played a leading role in such computer applications (see Figure 4.92). This clearly is one of the most rapidly developing fields at the physics–chemistry interface.

Chemical information, other than theory, also feeds back into physics, often in the form of specific information about substances having some particular desired properties. The ruby and neodymium lasers, the dye lasers, and liquid lasers in general were made possible in part because of the availability of the kind of information that the chemist now collects routinely. The subject of properties of materials also includes problems of preparation and purity. The development of solids for microcircuitry required new chemical research into methods of purification and analysis. Properly functioning microcircuit elements of one large computer were developed only when the physical and chemical problems of ion migration under conditions of high current flow and methods to achieve an exceedingly high degree of purification had all been studied in a quite fundamental way. (See section on condensed-matter physics.)

[255]

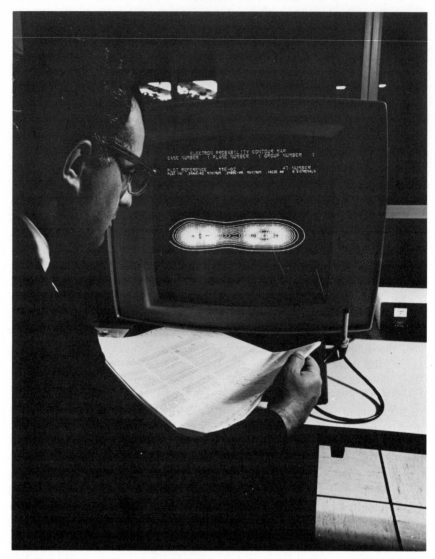

FIGURE 4.92   Computer-generated electron density map of the reaction $NH_3 + HCl$ → NH_4Cl, taken in a plane along the NCl axis, is projected on a display tube. [Courtesy J. D. Swalen, IBM Research Laboratory, San Jose, California.]

An example of an active area of materials preparation, in which chemistry and physics interact strongly, is the new field of bubble memory devices. It is hoped that these devices, based on the behavior of orthoferrite materials, will lead to a new generation of computer memories with far fewer failures and faster access time than are presently available.

One type of interaction between chemistry and physics falls outside the categories that we have mentioned. It relates to data compilation. The American Petroleum Institute has produced extensive tables of physical properties that, in effect, correlate properties of substances with their molecular structure. The National Bureau of Standards and the Joint Army, Navy, NASA, Air Force Interagency Propulsion Committee (JANAF) prepare similar compilations with other kinds of data. The National Bureau of Standards also has a system to provide standard samples for those who need them.

Fields Deriving Information from the Physics–Chemistry Interface

Many areas outside physics and chemistry receive benefit from a combination of ideas generated in these fields. In some cases the contributions take the form of very specific information; in others the contributions are more in the nature of general approaches and techniques of common currency in physics and chemistry that are just now being recognized as useful in other fields.

Engineering and technology are the most obvious fields in which the knowledge obtained from basic studies in chemistry and physics finds early application. Studies of corrosion, the synthesis of new materials such as crystalline polymers, the nature of surfaces, heterogeneous catalysis including enzyme reactions, and thin films require concepts and methods from both physics and chemistry. The techniques employed, such as low-energy electron diffraction, electron microscopy, electron microprobes, and electron spectroscopy are being assimilated into chemistry and physics. An example will show the interaction between chemistry and physics and the close relationship to technological developments. A manufacturer recently developed a material for photocopying; the material is an organic charge-transfer photoconductor. The development of this material can be traced from the basic concepts of the quantum theory of the chemical bond through the chemical idea of a charge-transfer complex to the recognition of photoconductivity. Current knowledge in basic physics and chemistry permitted the research for this material to advance rapidly to the point at which it was apparent that an electron acceptor was desired, thus reducing the need for expensive trial-and-error research.

The study and management of the oceans, the earth, and the atmosphere rely heavily on both physics and chemistry. In oceanography, for example, the turnover of water in the oceans can be understood only through a study of both hydrodynamic flow (traditionally a subject of physics) and molecular flow (as much a part of chemistry as of physics). The carbon dioxide balance in the air and the absorption of carbon dioxide by the oceans are critically important and difficult basic problems in physics and chemistry, as well as in biology, meteorology, and environmental management. The management of fisheries and water resources generally requires inputs from physics, chemistry, and biology. Air resource management is perhaps one of the clearest examples in which chemistry and physics play an integral role. Whether one is concerned with control of dangerously acute pollution levels or with long-range planning and legislation, air-quality models are essential to the analysis. Such models require a knowledge of the behavior of masses of gas (fluid dynamics), the reactions within the gas (aerochemistry), and the influence of the earth's shape, motion, and position on the way this gas behaves (meteorology and geodesy). Only within the past two years have the first models begun to appear that incorporate all these components at a level of accuracy high enough to be useful for an air resource management program. One of the major drawbacks has been the limited knowledge of the basic physics and especially of the basic chemistry of polluted air.

In social or economic modeling, a particular kind of contribution that the physicist or chemist can make merits attention. The constraints that one puts into a model are, in fact, of two kinds, although models now in use do not make this distinction. One type of constraint is essential and unavoidable, depending only on the laws of thermodynamics and other equally general laws. The other type of constraint is basically a limitation due to the state of current technology. By sorting the first type of constraint from the second, one can sometimes determine the ultimate limiting behavior of a system—the condition it could approach (but never reach) if its technology were optimum. Such an analysis could be made, for example, to determine the optimum long-term choices for supplying sufficient water to an industrial nation. A similar kind of analysis, using population biology instead of thermodynamics, could be used to determine the amount and pattern of insecticide to apply in an area to maintain the long-term agricultural productivity and the health of the animal species in the associated food chain.

Some other fields of research or development to which physics and chemistry are contributing include: the structure of biomolecules, the process of photosynthesis, the structure of cell walls, and the action of nerves; new methods of materials preparation; discharge methods for petroleum cracking; surface curing by electron bombardment; geographics, including the origins of vulcanism, the chemical origin of rocks, and minerals, and the composition of the planets; and the upper atmosphere and astrophysics, including the entire complex set of problems associated with the earth's atmosphere, as well as the problem of the formation of molecules in interstellar space.

Manpower

Because the definition of the physics–chemistry interface is not precise, estimates of the number of persons working at the interface are only approximate. The total interface contains between 4000 and 8000 scientists at the PhD level. (The larger figure includes condensed-matter physics.) This estimate is based on figures from the 1970 National Register of Scientific and Technical Personnel and on an independent estimate derived from the number of subscribers to the *Journal of Chemical Physics*. In chemical physics alone, the number of scientists is some 3000 to 4000. Of these, approximately 250 hold academic positions in chemical physics, and approximately 500 hold academic positions in some other subfield included in the interface.

The 1968 National Register recorded 8466 scientists working in physical chemistry, 4836 of whom had PhD's. The Register category, physical chemistry, is composed of an aggregation of detailed specialties. A scientist is classified as working in a field on the basis of his own statement of a detailed specialty that most nearly corresponds to the subject matter with which he is principally concerned in his work.

Many chemical physicists, of course, are doing work in subfields outside the scope of the Register definition of physical chemistry and would be included in, for example, the condensed-matter or fluid-physics subfields.

According to the National Research Council Office of Scientific Personnel, there are about 500 PhD's produced each year in "Chemistry, Physical" (459 in 1968 and 554 in 1970). (These physical chemists are produced by chemistry departments; no record is kept of "Physics, Chemical" as a subcategory of physics department PhD production.) For purposes of comparison, it is worth noting that in 1970 physics departments produced 400 condensed-matter doctorates, 470 elementary-particle- and nuclear-physics doctorates, and 730 doctorates in 11 other subfields of physics, including 160 in "Physics, Other." The total number of physics PhD degrees reported in 1970 was 1600.

The throughput—the replacement rate required to keep existing positions filled—currently is estimated to be about 6 ± 1 percent per year in the interface. This figure is based on several estimates, such as the normal throughput in a typical industrial laboratory. The figure appears to be fairly firm, except in years such as 1970, when the rate of hiring was held at a lower level. Consequently, the total number of replacement personnel required at the PhD level is approximately 240–280 per year for the over-all interface and about 180–240 per year for the more strictly defined field of chemical physics.

Based on a detailed analysis of manpower needs and production of degrees in subjects at the physics–chemistry interface, the Panel on Physics in Chemistry concluded in its report (see Volume II) that the rate of production of trained manpower in chemical physics is much higher than can be assimilated through the traditional past employment patterns of chemical physicists.

Problems at the Interface

The effects of reductions in financial support for this interface are less obvious, perhaps, than in areas in which a few large facilities dominate the field. The interface is characterized by a large number of small groups; most of the academic groups consist of one senior scientist, perhaps one or two young postdoctoral fellows, and a few graduate students. In government and industrial laboratories, the groups constituting research units consist of one, two, or perhaps three PhD scientists with, at most, the same number of technicians.

Financial cutbacks have thus far apparently taken the forms of non-replacement of staff lost by natural attrition, releasing staff, and, in some cases, loss of contract or grant support. Insofar as the elimination of support could be used to maintain excellence in the interface, this method would perhaps be the least unpalatable means of absorbing cutbacks. As the Report of the Panel on Physics in Chemistry states in its recommendations, supporting a smaller amount of excellent work at an adequate level is preferable to supporting a considerably larger amount of excellent, mediocre, and low-quality work at a level too low for any group to do an adequate job.

It is not clear, however, that the means of reducing funds are actually preserving excellence. The continued reduction in research support from the DOD and AEC has reached the point at which some of the best research is now threatened with loss of funds, and it appears that the NSF will not be able to assume this support at the level necessary to continue the work, at least within the current NSF budget.

Several national laboratories have experienced difficult problems; it has become necessary to terminate the employment of many able scientists.

So large a number has been released—people respected highly by their peers—that morale in the physics–chemistry interface at a number of national laboratories is at a low level.

> Biology has become a mature science as it has become precise and quantifiable. The biologist is no less dependent upon his apparatus than the physicist. Yet the biologist does not use distinctively biological tools—he is always grateful to the physicists, chemists, and engineers who have provided the tools he has adapted to his trade.
>
> Until the laws of physics and chemistry had been elucidated it was not possible even to formulate the important penetrating questions concerning the nature of life.
>
> PHILIP HANDLER
>
> *Biology and the Future of Man* (1970)
> (Chapter I, pages 3 and 6)

PHYSICS IN BIOLOGY

Nature of the Interface

An especially active scientific interface, and one that is attracting an increasing number of physicists, is that between physics and biology. The problems posed relate to fundamental questions of life and hold the promise of substantial contributions to the alleviation of human ills and misery. This interface combines fundamental science and both immediate and long-range social goals in a close and balanced relationship. Growing application of the methods, devices, and concepts of physics to some of the central biological problems should result in increasingly rapid progress. The scientific returns from such applications are already impressive.

The title of this section, "Physics in Biology," rather than the more usual biophysics, is indicative of the Committee's approach to the interface. Rather than a survey of all biophysics, this report focuses on the role of physics and physicists in attacking some of the major problems of modern biology. The implications of such a role for physics education and the overall physics enterprise also receive attention.

To describe the interaction of physics with biology requires study of the flow of manpower, ideas, and procedures through the interface. The title, "Physics in Biology," implies a flow from left to right. It is convenient to divide the subject into three parts: (a) the flow of physicists into biology;

[261]

(b) the flow of physics into biology, which includes the flow of ideas, techniques, and equipment, often through the intermediaries of engineering and chemistry; and (c) the interface called biophysics, in which the powers of physics are merging with the problems of biology in new ways that require nonstandard combinations of skills from both disciplines.

The goal of biophysics is to be biologically useful. When it is successful, it merges with other branches of biology, such as biochemistry and molecular biology, a major goal of which is to understand, in molecular terms, how genetic information is transmitted. The following example illustrates the contributions of biophysics to molecular biology. In 1944, it was shown that DNA, rather than proteins, contained the genetic information with which molecular biology is principally concerned. In 1953, J. Watson and F. Crick, working in the Cavendish Laboratory, developed from the accumulated chemical studies of DNA and a consideration of its biological function an interpretation of the x-ray studies of M. Wilkins in terms of the now famous double helix. Since then, molecular biology has developed rapidly, and biological function is now so interwoven with the structure of DNA that the term "structure and function" has become a platitude.

At the same time, and in the same laboratory in Cambridge, the crystal structures of the proteins were first being determined by x-ray crystallography, with a parallel influence on enzymology and biochemistry. In all these cases the isolation of the molecules and the definition of their biological functions were accomplished after roughly a century of chemical research. The physicists who determined the crystal structures were attacking a biologically important problem. Their boldest and most original step was their starting assumption that these large biological molecules had unique structures that could be determined by x-ray crystallography. Physicists are accustomed to this kind of simplicity in science; assuming such simplicity is a reflection of their previous training and research style. The revolutionary nature of their findings depended on the originality of their assumptions. However, given their background as physicists, trained in x-ray crystallography under Bragg, and their interest in biology, their directions of attack were almost predetermined. Thus, in these illustrative and illustrious cases, physicists and physics created an exciting field of research in biophysics, which has now been merged with biochemistry and molecular biology.

Determination of the structure of large biological molecules will continue to be an active field of research. In addition to x-ray crystallography, other physical techniques such as nuclear magnetic resonance, electron spin resonance, Mössbauer studies, and optical studies are being used more and more. These techniques, when used for structural studies, often complement x-ray data by giving information on a finer scale. This research is directed toward structural determinations of larger molecular aggregates, that is to say, membranes and membrane-mediated enzyme systems; ribosomes, which are the site of protein synthesis, composed of nucleic acids and protein, with a molecular weight of $\sim 10^6$; mitochondria, the membrane-bound volume in which the chemical energy of nutrients is converted to more usable forms by electron transfer reactions; and the photosynthetic unit in which photons are converted to chemical energy. In all of these systems, scientists are trying to understand biochemical functions in terms of the structure of the molecules and the physical interactions among them. Beyond the structures of isolated biological molecules lie the complicated questions of intermolecular interactions, which should challenge physical methods for many years. The collision techniques, which are only now beginning to be applied to the elucidation of elementary chemical reaction kinetics of simple inorganic molecular complexes, were developed in atomic and nuclear physics. Application of these approaches to molecular systems of biological interest is an exceedingly difficult but highly promising field.

When structure is examined at finer levels than the molecular, it is quite clear that quantum-mechanical understanding of the electronic structure of certain parts of biological molecules will become increasingly important. The advances made through electronic understanding of the molecules of interest to chemists and condensed-matter physicists show the promise of this approach. Recently, as experimental molecular physicists have studied biological molecules with the goal of understanding their electronic properties, the amount of systematic data has approached the point needed for theoretical synthesis and advances. This synthesis could lead in the future to a larger role for theoretical studies. Previously, the theorist's contribution to molecular biophysics has been very small, because, unlike the best experimenters, he generally has not learned enough biology to be able to ask good questions.

An exception occurs in the case of the theoretical models that are playing an increasingly important role in biology. This trend reflects the physicist's typically different viewpoint on biological problems. One of these differences is the physicist's desire for a simple, comprehensive model, capable of providing a first-order explanation of a wide variety of observations. It is often baffling to a physicist when biologists insist on the complexity of nature and the uniqueness of each result. It is, of course, equally unappealing to a biologist, struggling with complexities of DNA replication, to be informed by a physicist that the Ising model, or enough molecular quantum mechanics, would solve his problem. However, in the middle ground between these two extremes of oversimplification lies the productive application of physical models, based as always on experimental observations. For example, the concept of a genetic code proposed by physicists and theoretical chemists such as Crick, Orgel, Gamow, and Griffith appealed to a mind trained in physics. Various theoretical models were examined. One question that was proposed and answered was how much information had to be stored. The answer was that there were 20 amino acids that had to be coded by the DNA. Because DNA has only four possible bases as coding units, a minimum of three bases is required. Was the code overlapping? This question was answered negatively by considering the known mutations that had been observed.

Another useful model was that of Monod, Wyman, and Changeux, on allosteric proteins, whose function with respect to one small molecule can be affected by other small molecules. They proposed a generalized molecular basis for feedback in biological molecules and thus stimulated many experiments and analyses to determine the crucial facts.

Activity

Perhaps the most active research area at the interface between physics and biology is that involving the study and determination of the molecular bases for biophysical processes. This work has engaged some of the best people in the subfield and uses a variety of physical techniques and probes, from x-ray crystallography through nuclear magnetic resonance and Mössbauer techniques (see Figure 4.93) to nanosecond fluorimetry.

An older research area, but one that retains excitement and interest, is neural physiology. In part, this interest reflects the hope that such research can lead eventually to the understanding of the mysterious processes of human thought and memory, one of the remaining frontiers of man's understanding. In part, it reflects the physicist's assumption that when information is transmitted and processed by essentially electrical mechanisms, the problem should be amenable to physical analysis.

A striking example of physical reasoning in elucidating a particular property of a biological cell is the analysis of the electrical state underlying excitability in the giant axon of the squid. Its virtually unique diameter (500–100 μm) enabled Hodgkin and Huxley, in 1949, to conduct a series of fundamental electrical measurements that, in turn, made it possible to establish for the first time an adequate quantitative description of the electrical state associated with the nerve impulse. Both the design and execution of the experiments required a thorough knowledge of electronic circuits, in which feedback plays a crucial role, and of the theory of ionic electric currents. Moreover, the interpretation of the data demanded an ingenious mathematical analysis. This achievement was in large measure a product of Hodgkin's and Huxley's training in physics.

One of the basic findings of their analysis was that the ionic currents in

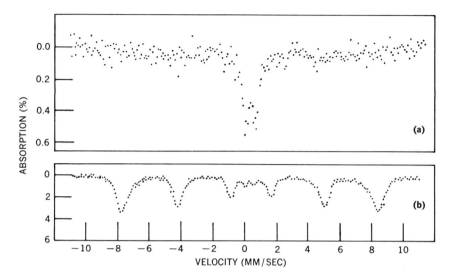

FIGURE 4.93 Human lung material from healthy lung (a) and from lung of hemosiderosis victim (b). Mössbauer spectrum of diseased lung indicates an abnormally large amount of iron (note the difference in absorption scales in the spectra), which appears to be in the form of a finely divided, low-molecular-weight compound. [Source: C. E. Johnson, "Mössbauer Spectroscopy and Biophysics," Physics Today *24*, 40 (Feb. 1971).]

the axonal membrane display strikingly nonlinear behavior in that the conductances are voltage-dependent and time-variant. The property of nonlinearity implies that, in a single neuron and chains of neurons (neural circuits), the elaborate calculations required to treat the excitable state can be carried out in general only with modern computers. And the statistical physics underlying the conductance changes in the cell membrane and at the junction between two cells (the synapse) presents a problem requiring highly sophisticated analysis.

The study of neural physiology is typical of the advanced research conducted on macromolecular aggregates in modern biophysics. It involves the design and development of new measurement techniques, computer simulation of neural behavior, and the study of signal-transmission characteristics of biological media. It continues to provide important surprises. Recent work on the brains of primitive animals, in which the brain contains at most only a few hundred cells, has shown that, even here, a remarkable symmetry of structure and function has developed.

Neural physiology stimulated some of the earliest physicists to move into biophysics; probably it will continue to attract them. Progress in neurobiology will demand advances in the biochemistry and ultrastructure of the neuron; but, in any event, the elucidation of physical mechanisms, for example, the analysis of excitability in the giant axon of the squid, will continue to play a crucial role.

A third major activity at the interface involves the interaction of radiation with high- and low-level biological systems. The types of radiation employed range from ultraviolet light to very-high-energy, heavy nuclear particles and mesons. The transfer of the techniques of nuclear physics— radioactive tracers, accelerator radiations, and nuclear instrumentation— has brought about a revolution in biophysics and in both clinical and research medicine.

Radioactive isotopes have contributed enormously to the general improvement in diagnosis, and a large number of radioisotopes are now in routine use. Isotopes commonly used include: ^{131}I, ^{125}I, ^{59}Fe, ^{113m}In, ^{99m}Tc, ^{51}Cr, ^{57}Co, ^{60}Co, ^{75}Se, ^{85}Sr, ^{197}Hg, ^{32}P, and ^{198}Au. These are used for visualization of the thyroid, brain, liver, lung, kidney, pancreas, spleen, heart, bone, and placenta and for a variety of physiological tests in which the rate of disappearance or rate of uptake of a particular labeled substance reflects the function of a given organ system (see Figures 4.94 and 4.95). In 1968, there were some four million administrations of labeled compounds to patients, and the rate of use has been growing rapidly. These isotopes are employed routinely in practically every hospital in the United States.

Increasing attention is being given to the use of very-short-lived isotopes, particularly ^{11}C, which should be especially useful because of the enormous potential for incorporation into a wide variety of biological compounds, with consequent extension of the range of diagnostic procedures. The short half-life, some 20 min, markedly limits the time for synthesis of the isotope into the desired compound and the time available for use. Thus the source of the isotope (an accelerator), facilities for rapid synthesis, and clinical facilities must be in close juxtaposition. A closely cooperating team of physicists, chemists, and clinicians is required for effective application. Use of this isotope, although still in its relatively early stages, is increasing rapidly and has great promise.

Another rapidly developing diagnostic procedure involves the determination of the entire amount of a given element, for example, calcium, in the body by means of activation analysis. The entire body is exposed to a beam of fast neutrons, which thermalize in tissue and are captured by the element in question. The patient is then placed in a whole-body counter and the total amount of the given element deduced from the total induced activity. The entire procedure can be accomplished with the delivery of only a small fraction of 1 rad to the patient.

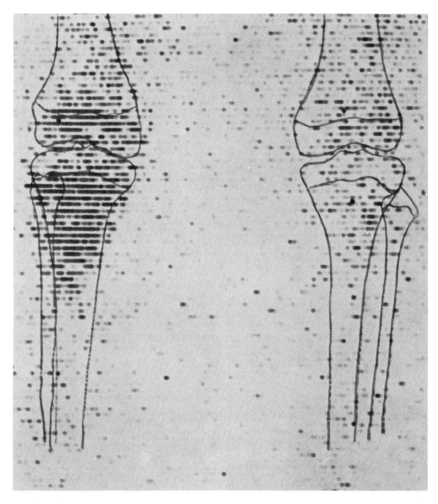

FIGURE 4.94   This is a bone scan made with short-lived strontium-$87m$ (half-life 2.8 h). The patient was a 13-year-old girl with a bone sarcoma of the right tibia. The greater strontium uptake in the right leg indicates the presence of the lesion. [Source: J. H. Lawrence, B. Manowitz, B. S. Loeb, *Radioisotopes and Radiation* (Dover, New York, 1969).]

Isotopes are now used widely in radiotherapy in a variety of ways. High-intensity external sources, such as cobalt-60 and cesium-137, have in many instances replaced x-ray machines for routine radiotherapy. The depth–dose characteristics of radiations from these sources are more favorable than those of most x-ray beams, and the units have the advantage of

[267]

FIGURE 4.95 *Left:* The autofluoro-
scope detector shown with its 2-in. lead
shield removed. A bank of 293 sodium
iodide crystals is in the lead-encased enclosure at bottom. This bank is separated
from the 12 photomultiplier tubes by a
4-in. Lucite light pipe. The data are
transferred electronically and recorded on
Polaroid film.

Below: Four neck scintiphotos of different individuals made with the gamma-
ray scintillation camera 24 h after administration of 25 to 50 μCi of iodine-131.
Upper left: Normal, butterfly-shaped,
thyroid gland. *Upper right:* Solitary toxic
nodule in right lobe. This "hot" nodule
takes up the iodine-131 to a greater extent
than the normal thyroid tissue. *Lower left:*
Degenerating cyst seen as a dark area in
lower left of picture. This "cold" nodule
is nonfunctioning, hence does not take up
the radioisotope. *Lower right:* This patient
had undergone a "total" thyroidectomy 2 years previously. The photo shows regrowth
of functioning tissue, right, and also a metastasis, lower center.

[Source: J. H. Lawrence, B. Manowitz, and B. S. Loeb, *Radioisotopes and Radiation* (Dover, New York, 1969).]

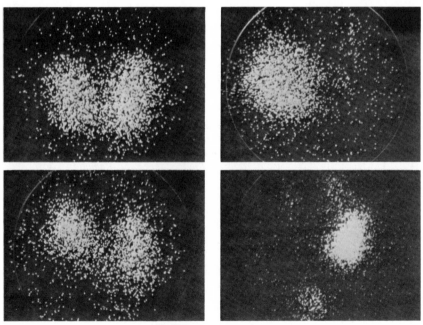

ease of operation and maintenance. As a specific example, the cure rate in cases of mammary cancer increased dramatically when cobalt-60 therapy replaced that with 100-kV x rays. The reason was that the higher energy ^{60}Co radiations were able to penetrate the sternum to a lymph node behind it, whereas the x rays could not. In addition, beta emitters such as strontium-90 are beginning to be used more frequently for the therapy of some superficial external lesions.

Physiological localization of radioactive isotopes is used in some forms of therapy, for example, iodine for treatment of hyperthymism and thyroid tumors and ^{32}P for treatment of some diseases of the bone marrow. These procedures represent optimal therapy only in a relatively few situations.

Accelerators have contributed greatly to the improvement of radiotherapy. Early accelerators allowed transition from the use of relatively low-energy x rays for radiotherapy to the use of supervoltage x rays, permitting the delivery of a relatively large dose to the tumor in depth, with a minimal dose to the intervening normal tissues. Electron accelerators such as the betatron have permitted an additional distinct improvement in the therapeutic ratio, or dose-to-tumor, dose-to-normal-tissue ratio.

Somewhat in the future is the therapeutic use of beams of negative pions, which currently are produced only in elementary-particle and very-high-energy nuclear-physics facilities. These have the enormous advantage of delivering not only their ionization energy but also their entire rest mass energy to their final destination in matter. Thus, while traveling to the therapy site, they do relatively little damage to surrounding tissues; their capture then releases some 200 MeV of energy at the treatment site.

Currently, there is much interest in the use of accelerators to produce beams of fast neutrons for radiotherapy. The rationale is that all tumors quickly develop small foci of poorly oxygenated or hypoxic cells. These hypoxic cells are markedly resistant to damage by x or gamma radiation but are much more susceptible to damage by neutrons or other densely ionizing radiations. Although a variety of reactions and neutron spectra might be used, the approximately 14-MeV neutrons from the D–T reaction are optimal in terms of penetrating characteristic and density of ionization. The procedure is experimental, and several years will be required to evaluate its efficacy. Man-made transuranium isotopic sources of neutrons, such as ^{252}Cf, have just recently become available and are also being used, with the same rationale.

Not only are radiations in the high-energy or nuclear realm used in such studies and applications but also ultraviolet and infrared radiation (see Figure 4.96). For example, Setlow and his collaborators at Oak Ridge very recently discovered that a particular species of skin cancer can be traced not only to a specific damage site in the human cell DNA induced by

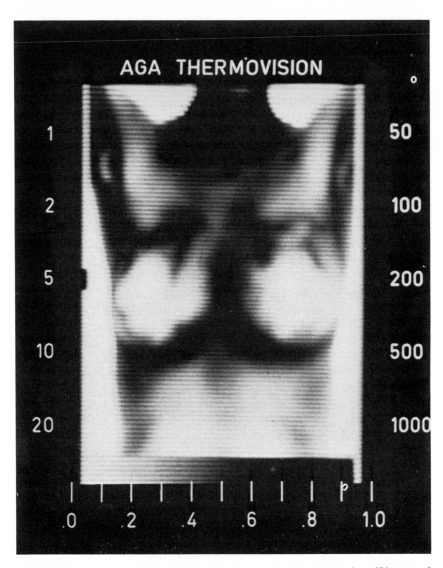

FIGURE 4.96 Detection of breast cancer by infrared thermography. [Photograph courtesy of the Lovelace Foundation for Medical Education and Research.]

ultraviolet radiation but also to the lack, in susceptible individuals, of a rather rare enzyme the function of which is to repair such damage. Already a test has been developed that can detect the total lack of this enzyme *in utero* and make possible therapeutic termination of the pregnancy; otherwise, with total lack of the enzyme, the normal life-span of a child would be brief. It is hoped that further research will result in a technique for the detection of, and compensation for, the partial lack of this enzyme in the population.

The use of infrared photography as a diagnostic tool is now rather well developed. Reflecting the increasing metabolism of cancer cells is a local temperature elevation that shows clearly in an infrared photograph; this technique is extremely simple, over 90 percent effective, and widely employed.

The long-term effects of very-low-level radiation on a population is, of course, a matter of continuing concern and controversy. The controversy arises largely because, at the levels now under consideration, even granting the validity of the mouse–man extrapolation, it has been estimated that an 8-billion-mouse colony would be required to yield statistically significant results. This situation shows the importance of understanding mechanisms not just doing statistical experiments. It is an example of a situation that Weinberg has defined as "trans-science," * a problem in which it appears superficially that in principle science should be able to give concrete answers but, when examined in greater detail, exceeds the scope of any economically feasible scientific study.

Although now of somewhat less importance than in past decades, the overall questions of thermodynamic energy balance and stability in biological systems continue to occupy the attention of a small group of physicists in biology. Although the broad outlines of the physics involved in the intricate energy-transfer mechanisms have been established, major questions remain unanswered.

In fluid physics and rheology there also are problems. The dynamics of human body fluids are still inadequately understood and can pose significant problems in vascular surgery and repair and in the more ambitious organ-replacement projects now in developmental phases. In rheology, a wide range of open questions in regard to bone growth, muscle attachment, and the like need answers.

Biomedical engineering has become a specialty in its own right, stimulated by the pressures for improved man–machine interface designs in supersonic aircraft, space vehicles, and precision industrial production lines, as well as by more prosaic needs such as improved kitchen appliances. The

---

* *Science, 174,* 546–547 (Nov. 5, 1971).

design of artificial organs, prosthetic devices, and the like is another part of this field. Progress in the development of long-lived portable power sources and parallel progress in ultraminiaturization of semiconductor electronic components should lead to a greatly increased capability to mitigate human infirmities. Much of this work represents applied physics at its best.

One of the most difficult tasks in the biomedical engineering areas of bacterial colony analysis, brain scintigram analysis, blood-cell identification, chromosome analysis, and heart image extraction from a cardioangiogram is the extraction of relevant objects from an irrelevant background. The principal reason is that the pictures in these fields are often complicated by unwanted background, and object images are poorly defined. In recent years, computer processing of radiographic images has emerged as a highly promising technique (see Figures 4.97 and 4.98).

Three aspects of biophysical instrumentation merit attention. The first involves instrumentation for clinical application in both diagnosis and treatment. Major progress is under way in clinical instrumentation; a modern hospital's intensive-care wing illustrates the vital role that physics research plays here. A second aspect of instrumentation involves biological and biochemical laboratories in which new techniques, devices, and approaches permit massive increases in both the speed and quality of measurement, thus making possible the extension of the most modern diagnostic aids to a much larger segment of the population. The techniques also facilitate ongoing research in biology, biochemistry, and biophysics. Third, advances in instrumentation open entirely new areas of biophysical research. Examples of devices that have led to major breakthroughs are the scanning electron microscope (see Figures 4.99 and 4.100), the ultracentrifuge, and nuclear magnetic resonance.

Institutions

Research at this interface takes place in universities and in a number of large federally supported laboratories or institutes. The principal supporting agencies are the National Institutes of Health (NIH) and the AEC. Some more specialized work is supported in-house by NASA and the DOD. A relatively small fraction of the total effort is in industrial organizations. The division between university and federal laboratory activity is roughly equal.

Academic research at this interface, in contrast to research at institutes, carries a heavy burden of departmental responsibilities. Two means have been used to broaden the scope of academic research. In many cases, temporary departments of biophysics were established and made into formal departments when it seemed desirable. This practice has worked well in a number of universities and less well in others. In the most successful cases,

FIGURE 4.97 Quantitative information to assess the functional state of the heart, especially the left ventricle, can be obtained from a cardiac cineangiogram—an x-ray motion picture of the heart. To collect such information, an essential task is to detect or outline the boundary of the left ventricular chamber on each frame of the cineangiogram—a laborious task, heretofore performed by humans. C. K. Chow and T. Kaneko (IBM, Yorktown Heights, New York) have developed a computer algorithm to detect the left ventricles automatically; thus making the automatic extraction of quantitative information feasible. *Above:* The scanned image of one frame of a cardiac cineangiogram; *below:* the boundary detected by the algorithm superimposed on the image. The detected boundary agrees reasonably well with human recognition.

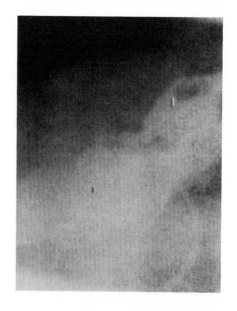

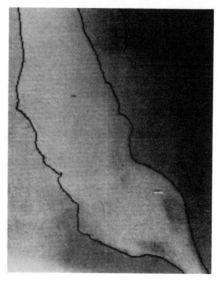

such biophysics departments became departments of modern biology and often have lost much of their initial chemical or physical orientation. However, the more physics-oriented aspects of biophysics still flourish in university departments of this sort, generally under a title such as "Biophysics and Molecular Biology." Advanced electron microscopy research, new

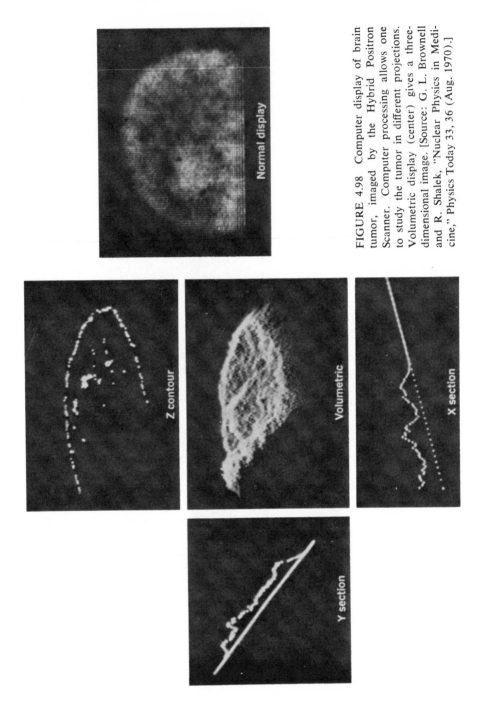

FIGURE 4.98 Computer display of brain tumor, imaged by the Hybrid Positron Scanner. Computer processing allows one to study the tumor in different projections. Volumetric display (center) gives a three-dimensional image. [Source: G. L. Brownell and R. Shalek, "Nuclear Physics in Medicine," Physics Today 33, 36 (Aug. 1970).]

Normal display

Z contour

Volumetric

X section

Y section

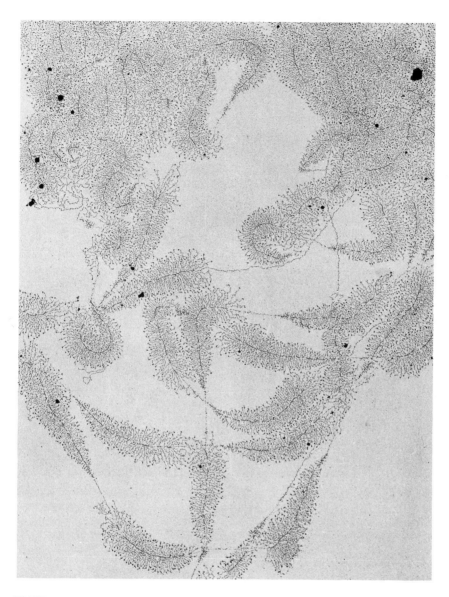

FIGURE 4.99 Photographs of genes in action were taken for the first time in 1969 at Oak Ridge National Laboratory in Tennessee. Single genes on the long central strands of DNA have thousands of shorter RNA strands peeling off of them. [Source: *Science Year. The World Book Science Annual.* Copyright © 1969, Field Enterprises Educational Corporation.]

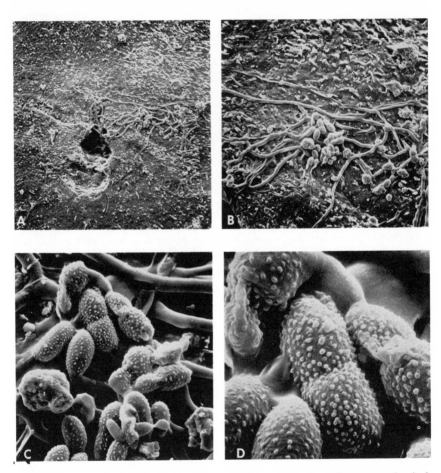

FIGURE 4.100   Surface of a Sturmer apple fungus and spores adjacent to a lenticel. Some hyphae appear to enter the lenticel, possibly indicating first stage of penetration into the apple. Part of a study concerned with the occurrence of lenticel spot. (a) Magnification ×160; (b) magnification ×400; (c) magnification ×1600; (d) magnification ×12,000. [Photographs courtesy of Stereoscan Micrograph—Cambridge Scientific Instruments Limited, England.]

techniques of radioactive tracers, magnetic resonance studies, and x-ray crystallography have prospered in these departments and profited from their interactions with the more biologically oriented activities. In practice, these departments do not interact to a large extent with physics deparments, although physics courses are increasingly important to their students. In this respect, it is generally felt that physics departments do less than chemistry departments to accommodate such students, among whom are an increasing number of premedical students who desire some physics courses. Revision of the physics curricula to accommodate the needs of these students would be one way in which the interactions of physics and the life sciences could be strengthened.

There are important opportuntities to further these interactions in the large national laboratories initially established for specific mission purposes. These laboratories have extensive facilities and trained staffs who could bring a unique competence and capability to bear on problems at the physics–biology interface. A pressure that is favoring larger research groups is the increasing cost of the equipment used in biophysical research. The cost of commercially available high-resolution nuclear magnetic resonance (NMR) equipment used to measure the proton NMR spectra of complicated organic molecules can be cited as an example. In 1957, this technique was first used to study the spectrum of a protein. During the past decade, stimulated in part by the possibility of extensive biological applications, this equipment has increased rapidly in sophistication and price. At first, the cost increase followed a semilog growth curve, but recently, because of the use of superconducting solenoids and Fourier transform computer techniques, the costs have been increasing more rapidly with time (see Figure 4.101). At the same time that the equipment is becoming too expensive for an individual scientist to afford, its usefulness in biochemical research has been increasing rapidly because of improvements in sensitivity, resolution, and experience. Clearly, large institutional research activities, similar to the national laboratories, will soon be more useful in biophysical research.

Manpower

The number of people identified with physics in biology, or biophysics, depends on how the interface is defined. The number of physics participants in the 1970 National Register of Scientific and Technical Personnel who indicated physics in biology as the subfield in which they were working was 465, or 1.3 percent of the 36,336 physicists who took part in the survey. Sixty percent (277) of the 465 held PhD's, representing 1.7 percent of all physics PhD's.

If the population of the interface is based on membership in the Biophysical Society, then it increases to approximately 2500. The Society was founded in 1957, and its growth has been extremely rapid.

The numbers of positions and graduates seem to be reasonably well balanced in this interface between physics and biology. There have been fewer positions in the past few years, as in all of physics, but the situation seems no worse in physics in biology than in physics in general.

The contributions of physics are essential to modern biological research

[277]

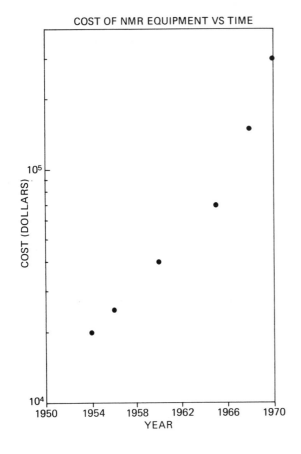

COST OF NMR EQUIPMENT VS TIME

FIGURE 4.101 Increases in the cost of nuclear magnetic resonance equipment, 1950–1970.

and are likely to increase rapidly. Research in biophysics is particularly rewarding in a personal sense, first, because of the exhilarating intellectual challenge of this wide open, diversified, and fast-moving field and, second, because of the humanistic aspects of biological research, with its long-range goal of improving human life. In the future, many more physicists probably will move into biology. In the Report of the Panel on Physics in Biology (see Volume II), a number of ways in which this movement can be facilitated and encouraged are suggested.

[T]he only certain means and instruments to discover and anatomize nature's occult and central operations; which are found out by laborious tryals, manual operations, assiduous observations and the like.

JOHN WEBSTER (1610–1682)
Academiarum Examen, 1653 (page 106)

INSTRUMENTATION

Introduction

High-quality observations are the lifeblood of science; physics, in particular, because of its focus on fundamental causes, is critically dependent on the instrumentation that makes such observations possible. At the same time, physics provides a great many of the concepts and devices that underlie all modern instrumentation.

In surveys such as this one there is always a tendency to focus attention on the instrumentation applications of the most recent and spectacular scientific findings—for example, the laser—and on the instrumentation required to reach the frontiers of modern scientific research—for example, superconducting magnets in elementary-particle physics. Although these applications are very important, such a focus does a great disservice to the vital role of instrumentation throughout science, society, and the economy. Indeed, instrumentation is that channel through which the results of physics research are most directly brought to bear on the problems of greatest concern to the citizen.

In its survey the Committee attempted to avoid this too narrow view of instrumentation through the appointment of a panel of distinguished representatives of leading companies in the U.S. instrumentation industry. These companies and their products span the entire range from the most practical instrumentation used in heavy industry—for example, steelmaking monitors—to instruments of great delicacy and precision that are used in science and medicine—for example, atomic clocks and neurosurgical probes (see Figures 4.102–4.105). This section draws heavily on the report of this Panel on Instrumentation, which appears in Volume II. After presenting a brief historical background, we examine the role of instrumentation in both the scientific and economic life of the nation and draw certain comparisons with activities in selected foreign countries. This examination is followed by a brief discussion of the development process in the instrumentation industry, tracing, in selected cases, this development from basic physics through technical development to practical application. The section concludes with an examination of certain outstanding problems that the instrumentation industry faces at present.

FIGURE 4.102 Photograph of a modern airborne proton magnetometer used for making magnetic surveys, which are particularly useful to petroleum geologists in helping to identify formations in which oil might be found. [Photograph courtesy of Varian Associates.]

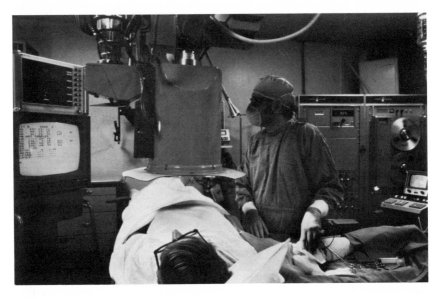

FIGURE 4.103 Computerized cardiac catheterization systems aid doctors in the prompt diagnosis and treatment of critically ill heart patients. [Photograph courtesy of Hewlett-Packard Company.]

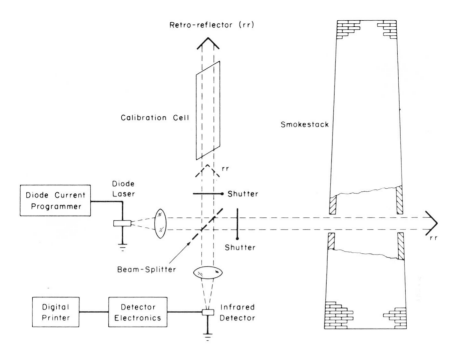

DIODE LASER SYSTEM FOR ACROSS-THE-STACK MONITORING

FIGURE 4.104   Diode laser system for across-the-stack monitoring of sulfur dioxide. This system is under construction for the Environmental Protection Agency and is based on the absorption of tunable laser radiation—yielding an average value for the pollutant concentration over the stack diameter. [Photography courtesy of Massachusetts Institute of Technology, Lincoln Laboratory.]

Applications of Instrumentation

Although there are a great many possible categorizations of instrumentation depending on use, precision, scope, underlying mechanisms, and the like, here we consider only three broad classifications. These are

1. Instruments for research, laboratory measurement, testing;
2. Instruments for industrial process management and control;
3. Instruments for analysis, which include types for both research and laboratory use and industrial plant on-line use.

Before examining these categories in detail, however, it may be useful to consider the applications of instrumentation more generally. A society

FIGURE 4.105   Lone Star Cement Control Center, Greencastle. [Photograph courtesy of Leeds & Northrup Company.]

without numbers, and, hence, a society without instruments, is inconceivable in today's world. There are very few types of human activity in which instruments are not necessary. Four activities involving major applications are the following.

Research   Instrumentation is a key and indispensable element in research, whether the research be attempts to advance man's knowledge of nature's laws, of the characteristics of nature's materials and energies or of life and biological processes; or to synthesize new materials or develop new techniques of energy conversion or new products and processes. Each major step forward in research creates a need for new and better instruments of greater resolution, capability, or flexibility, so that the next significant step forward in research can be undertaken and achieved.

The relationship between research and technology, as exemplified in the instrumentation and apparatus of research, has been essentially one of symbiosis. Advances in science have both spurred and made possible new technological advances; these, in turn, have made possible major advances in science. Examples of this symbiosis abound in this Report and in the panel reports in Volume II, and we return to this subject in a subsequent part of this section.

Industrial Processing   In the scientific and industrial sense, ours is a well-ordered world. Fortunately, physical and chemical properties of things or materials can be established by investigation and, under comparable conditions, can be repeated. The essence of production is to create materials and things—and in some cases to provide services—with fixed, predictable qualities, qualities that can be reproduced if the conditions and environment of production are properly established and maintained.

It is the function of instrumentation first to determine the magnitude of parameters that define the conditions of a process. Then, by automatic control means, incorporating in recent times computers of broad capability, appropriate variables are adjusted so that desired process conditions are achieved. Analysis instrumentation, either in the laboratory or on-line, depending on the prevailing state of the art, assists in regulating the process to yield products of desired quality while achieving economical process operation. The automation and process control that modern instrumentation has made possible have already had a profound impact on the quality of life throughout the world. The grinding monotony and drudgery that frequently characterized industrial working conditions for unskilled labor in the past are now largely gone, and, at the same time, the reliability and economic competitiveness of U.S. industrial products have increased significantly.

World Competition   Successful competition in the trade markets of the world is an essential element in maintaining a nation's standard of living and a correspondingly highly cultured life style. World competition is both technological and economic. Given steady trends toward equalization of basic technology, at least in some major fields, the competition in such fields resolves into economic comparisons, that is to say, largely comparative process and plant productivity.

The importance of productivity in world competition, in the face of other international economic trends, is illustrated in Figures 4.106–4.109. Figure 4.106 shows comparative wage increases over the past several years in the principal industrial producing countries of the Western world. Japan has had the highest rate of wage increases, with West Germany close behind. The United States, surprisingly, has had the lowest. Figure 4.107 provides an index to changes in consumer prices. Again, Japan is the highest, with

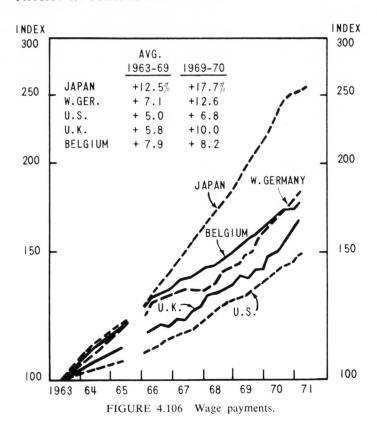

FIGURE 4.106   Wage payments.

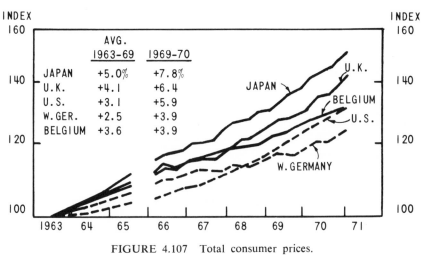

FIGURE 4.107   Total consumer prices.

[284]

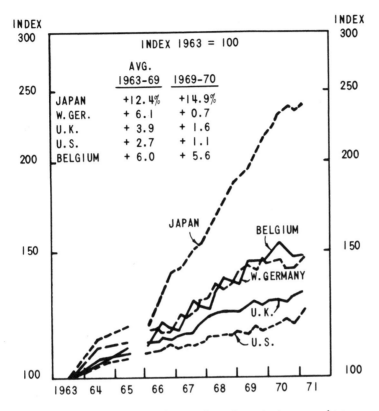

FIGURE 4.108   Changes in manufacturing output per man-hour.

the United States appearing between it and West Germany. Figure 4.108 then illustrates changes in manufacturing output per man-hour, which is to say, productivity. Japan shows the greatest improvement; the United States, the least. Figure 4.109 shows the net result of changes in wage rates and productivity, in terms of unit labor costs. It will be seen that Japan, despite the highest rate of wage increase, has had sufficient counterbalancing improvement in productivity so that its unit labor costs have been held substantially constant over the past few years. The United States, together with the United Kingdom, shows the highest increases in unit labor costs.

These curves demonstrate, quite emphatically, that U.S. productivity must be improved if this nation is to maintain a satisfactory position in world competition. Manifestly, the problem is not a simple one. A solution requires participation by, and significant contributions from, many sectors and groups. Advances in science and technology and related improvements

[285]

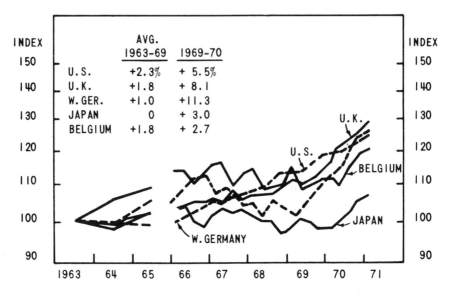

FIGURE 4.109   Changes in unit labor costs.

in process control by means of instrumentation can be important factors in achieving the requisite productivity improvements in U.S. industries.

Society   Throughout this Report and those of the panels we have emphasized the importance of the contributions that physics has made, and can make, to society. These range from the cultural and educational—both closely related to the extension of man's intellectual horizons in research—to the alleviation of human ills and suffering, through clinical and research medicine, to the satisfaction of material needs and desires of the nation's consumers, and to the assurance of safety and defense both locally and internationally. In all these areas instrumentation plays a vital role in the interface between the scientific principles involved and the actual application or device.

The extent of these contributions is great, but their existence and their benefits are difficult for the public to sense or see. One reason is that they frequently occur within the technical community and the capital equipment industry at stages where no consumer products or services are directly or visibly involved. Another is that the benefits are diffuse, composed generally of an extremely large number of individual improvements, efficiencies, and advances in an extremely wide variety of industries, institutions, and human activities. But they are there.

[286]

The U.S. Instrument Industry

Historical Background During the nineteenth century, instruments used in the United States were for the most part imported from abroad, primarily from Germany and England. At the turn of the century and in the decade that followed, domestic instrument companies were organized and developed. Also in 1901, the National Bureau of Standards was organized, and its pioneering work in the development of standards and measuring techniques profoundly influenced and stimulated the design, fabrication, and use of domestic precision and industrial instruments.

Thus, when World War I began, apparatus, standards, and techniques for measurement and control were available domestically in support of the nation's industrial and defense needs.

In the post-Depression years preceding World War II, a number of small companies, largely oriented to the emerging science of electronics, entered the instrument field. They, and other new groups, supplemented the efforts of the older instrument companies in providing major support to the nation's technical effort—in both research and production—during the Second World War. The technologies emerging from that war, including those based on the new nuclear science and on radar activity, found expanded use in world science and industry and stimulated considerable growth of U.S. instrument enterprises. Later, space programs, expanded defense needs, extensive government-sponsored research, expansion in industrial processing, and major scientific advances in solid-state physics and computer technology all contributed to new and larger needs for instrumentation, with corresponding growth in the U.S. instrument industry and—at least until the present time—worldwide predominance in this field. Until very recently, the U.S. scientist or engineer who visited laboratories in all parts of the world, except the Soviet Union or China, found that the instrumentation in use was familiar—in many cases identical to that in use in the United States and produced by the same manufacturer. This situation is changing rapidly to one of lesser dependence on U.S. instrumentation sources.

Present Instrumentation Industry Activity Four aspects of U.S. instrumentation activity merit particular attention.

1. *Size:* The current annual sales of the U.S. instrument industry are estimated at about $3 billion and are made by approximately 1800 establishments having over 165,000 employees. Company sizes vary widely: the largest has annual sales of about $350 million; numerous companies have annual sales in the range from $500,000 to $1 million. Small companies compete successfully with the larger ones on the basis of innova-

tive technology and aggressive applications activity in more narrowly selected product lines.

2. *Research and Development Activity:* Product lines are characterized by high technology content, and, although research and development activity elsewhere represents an important market for the industry, it ranks high among industries in the amount of sales reinvested in its own research and development. For individual instrument companies, the range of investments in new product development extends from 5 percent to 10 percent of sales. Leading companies are at the high end of this range. In 1969, the overall average for the instrument industry was 6 percent, with about 13,200 scientists and engineers employed in new product development activity at the end of that year.

3. *Contributions to the Balance of Trade:* World leadership in instrumentation technology coupled with sound business practices have enabled the U.S. instrument industry to achieve a high level of export sales. They amount to as much as 50 percent of the total annual sales in some products and average about 20 percent of the total annual sales for the industry as a whole. Exports of instrument products have exceeded imports in the ratio of 3.5:1, thereby contributing significantly to the nation's balance of trade.

4. *Professional Organizations:* The Scientific Apparatus Makers Association, with headquarters in Washington, D.C., now in its fifty-second year, is the principal national trade association for the industry. The Instrument Society of America, with headquarters in Pittsburgh, which was founded 27 years ago, is a professional society devoted specifically to instrumentation science and technology. It has 20,000 members and is represented on the National Research Council.

New Product Development in the Instrument Industry

Funding The U.S. instrument industry is a strong investor of its own funds in new product development. As previously noted, the industry invests, on the average, about 6 percent of its annual sales in in-house research and development, although some of the more successful companies spend as much as 10 percent or more of annual sales in such efforts. It should be emphasized that these rates of expenditure refer to corporate funds and do not include any special projects that may be funded within some companies by government agencies.

Time Scales in New Product Development There is a well-developed progression of steps involved in the conversion of a basic scientific principle into an acceptable reliable instrument or other product. Research, development, engineering design, tooling, testing, manufacture, and marketing are the requisite elements of the overall development process, from recognition of a need to implementation in the marketplace (see Figure 4.110).

The overall time for moving through these in-house steps will vary from perhaps two to five years, depending on the complexity of the product, the degree of innovation that it incorporates, the nature of the environments in which it will be applied, the capabilities of the purchaser's personnel, and the policies of the manufacturer in regard to field failures or inadequacies of his products. Generally speaking, new products will have been thoroughly tested and will be marketed only after the achievement of successful results in the environments of their proposed use.

An example of a relatively simple new product that required a much longer time for successful development than might have been anticipated initially is the expendable immersion thermocouple for molten steel (discussed in greater detail later in this section). A relatively simple product in appearance, based fundamentally on well-known concepts of classical physics, it nevertheless required several years of intensive innovative design work and testing before fully reliable units, adapted to the rough impersonal treatment and hostile environment of a steel mill and retailing for less than $1 each in spite of their noble-metal content, were produced.

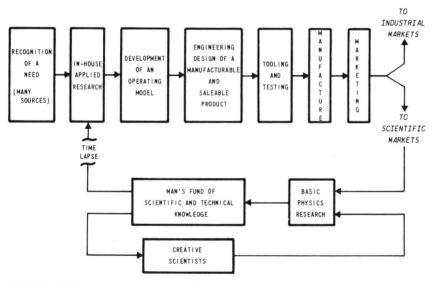

FIGURE 4.110   Steps and cycles in the development of new instruments.

The challenges to successful production and performance for such a device were great.

At the other end of the scale are highly complex analytical tools, such as nuclear magnetic resonance (which we discuss in greater detail later in this section), that, at least in their earlier period of introduction, are used in an atmosphere of sophistication by highly trained and sympathetic scientists or technicians who are conditioned to the uncertainties of research and are strongly motivated to have the equipment perform satisfactorily. Anticipating such an atmosphere of use may well encourage a manufacturer to reduce the development time required for initial versions of a product destined for these markets and anxiously awaited by potential users.

Risks in Product Development Frequently, the time span of successful product development proves to be much longer than anticipated, and the attendant development costs much higher than originally planned. Also, acceptance in the marketplace may be less than was expected. This lack of acceptance may result from unanticipated economic conditions, competitive breakthroughs and products, or overly optimistic estimates of what potential users really would be willing to buy. All these factors are risks in development, and, axiomatically, the greater the technical innovation, the greater the economic risk is likely to be. Nevertheless, equally axiomatically, the U.S. instrument industry is dedicated to taking these risks, recognizing that without them it cannot sustain growth and leadership in its selected markets.

As the industry performs its applied research in the hope of developing profitable new products, it has the comparable hope that in other appropriate areas there will be adequate support of basic research, for the results of such research will contribute to the industry, if not directly, then inevitably in time and in unanticipated ways. We have noted previously the great degree of symbiosis between the scientific and instrumentation communities. The research scientist is alternately discoverer, innovator, customer, and user, and in all these roles he advances the state of the art in the instrumentation industries. In all these roles, too, problems have developed, which we shall discuss.

Using the Results of Basic Physics Research

Possible Spinoff from Basic Research Creative scientists, basing their work on earlier discoveries, and frequently using instrumentation from commercial sources (as well as creating instruments of their own), engage in basic research, the results of which become significant additions to man's fund of scientific and technical knowledge. Such new knowledge is sought for its own sake as part of man's insatiable curiosity about his universe and his desire to penetrate and divine nature's mysteries and secrets. There is, in such basic research, no advance consideration of whether a practical use will be found for its results. But inevitably, and after varying lapses of time, such uses are found—often where least expected.

Because the instrument industry, like all other industries, is profit-oriented, its research and development programs have, with only rare exceptions, specific marketing objectives. Thus they are properly defined as applied rather than basic research. Basic research, in other words, is generally beyond the profit-oriented scope of instrument companies. Their research and development departments, however, do have an ongoing appetite for new knowledge. They are continually on the lookout for new scientific information available from other sources that might in time find a place in new products that are to be developed.

Examples In the Report of the Instrumentation Panel in Volume II, a large number of instrumentation systems deriving rather directly from basic physics research are listed, and case histories of selected ones are presented. We have selected two of these examples for purposes of illustration here. The first, the expendable thermocouple, represents an example of a simple device, based on classical physics, designed for use by unskilled personnel in an extremely hostile environment. The second, nuclear magnetic resonance, is an example of a sophisticated device based on quantum physics and designed for use by relatively skilled personnel in what might be characterized as a research environment. Despite the vast difference in these devices, however, they share the common attribute of making possible measurements that would be obtained otherwise only with great difficulty if at all and that have important economic and social consequences.

To manufacture steel economically, the steelmaker must know the temperature of the molten steel in the open-hearth, basic-oxygen, or electric-arc furnace so that the steel can be poured, or tapped, at the appropriate point. If the heat is tapped at too low a temperature, steel will solidify in the receiving ladle before it can be completely transferred to the ingots molds. This situation results in so-called skulls, which must be removed from the ladle, broken up, and remelted—obviously a wasteful procedure. If the steel is tapped at too high a temperature, it penetrates crevices in the mold, causing the ingot to adhere to its mold. This result is called a sticker. The mold must be broken to remove the sticker, again a costly procedure. Also, heating the steel to a temperature higher than is necessary lengthens the time of a heat and uses extra fuel, which increases the cost of the heat. In spite of intensive efforts over a period of five decades by steelmakers throughout the world, a fully satisfactory solution to the problem had not been developed by the mid-twentieth century.

Techniques tried during the 50 years of experimentation included thermocouples of various designs and radiation-responsive units. The high temperatures and hostile environment of the measurement militated against the attainment of successful results.

In 1958, a thermocouple unit sufficiently low in cost, despite the use of noble metals, to permit throwing it away after one use, and highly accurate in its single use, was introduced. It won rapid acceptance throughout the steel industry. It used a number of concepts derived from basic physics, although many of these concepts had been well known for decades. This device is an example of a lengthy time lapse before principles derived from basic physics were applied to practical instrumentation to fill a specific need. It emphasizes the challenges of in-house efforts to extend available knowledge effectively with innovative development and engineering activity to produce a reliable manufacturable product that can be sold at a profit.

Because the successful solution to the molten-steel temperature-measuring problem is a specially designed thermocouple used with a null-balance potentiometer recorder, it is of interest to sketch the historical background for these devices.

The thermocouple effect, that is, the development of a small dc voltage related to the temperature difference between the hot and cold junctions of two dissimilar metals, was discovered by Seebeck in Germany in 1821. At just about the same time, at Berlin University, Poggendorff devised the first known null-balance type of potentiometer, which he used to measure the output of electrochemical cells without drawing current from them. As far as is known, not then or for a considerable time thereafter, was there a common interest in or joint use of the thermocouple effect and the null-balance potentiometer. It took about 80 or 90 years for the two to be combined for industrial temperature measurement.

In about 1886, some 65 years after its discovery, practical application was made of Seebeck's thermocouple effect by Le Châtalier in France, who measured the output of the thermocouple with an indicating deflection millivoltmeter. Shortly thereafter, millivoltmeter-type recorders were adapted to temperature recording using thermocouple temperature detectors. But such recorders had limitations due to insensitivity and calibration drifts of millivoltmeter movements, the influence of length of leads on the temperature measurement, and difficulties related to reference junction compensation.

Meanwhile, along another path of temperature-measurement development, the first platinum resistance thermometer, depending on the temperature coefficient of platinum, was developed in Germany in about 1871. By 1890, a practical platinum resistance thermometer was developed in Great Britain.

In 1897, the Callendar Wheatstone bridge recorder, believed to be the first null-type recording instrument on record, was introduced. It used advanced concepts of electrical balancing but did not prove to be very practical in the hands of other than skilled operators.

Then, early in the twentieth century, individuals in industrial plants began to combine thermocouples with manually operated null-balance potentiometers to make precise measurements of temperature. This technique, which at balance drew no current from the thermocouple, had many clear advantages and stimulated efforts to develop an automatic null-balance potentiometer recorder to take the place of the manually operated instruments. By 1912, such a successful recorder was developed, and Seebeck's thermoelectric effect came into its own for industrial temperature measurement and control. Literally millions of thermocouples and tens of thousands of potentiometer recorders have since been applied to a wide variety of industrial processes. Nevertheless, despite the great success of this technique in other applications, it had not been applied successfully to the molten steel temperature-measuring problem by the mid-twentieth century.

Some of the problems associated with making a successful measurement of molten steel were the following:

1. The working temperature is 2900°F—high for an industrial measuring device.

2. An accuracy of 5 to 10°F is desired.

3. The molten steel is covered with a surface of slag.

4. Expensive radiation-responsive devices intended for multiple repetitive use require heavy protection, are cumbersome, and are subject to large errors.

5. Similarly, thermocouples intended for multiple repetitive use require heavy protection and limit the operator's ability to establish temperature trends before he has to tap the furnace. Also, repetitive exposure of the platinum–platinum, rhodium thermocouple to the high temperature of the bath causes significant calibration drifts.

What clearly was needed was a device that would overcome all these limitations and permit measurements to be made rapidly and accurately at a cost not exceeding $1 to $2 per measurement.

The unit introduced in 1958 that, together with a null-balance potentiometer recorder, successfully solved the molten-steel temperature-mea-

[293]

suring problem is a thermocouple of very special design. Its features are these:

1. It uses a platinum–platinum, rhodium thermocouple and achieves the 5–10°F accuracy by being used only once and then discarded.

2. It achieves the economic targets by using only a very small length of the precious thermocouple wires.

3. Base metals with suitable thermoelectric effects are used within the thermocouple in place of longer extensions of the precious wires. The junction of these base and noble metals must be at less than 200°F even while the thermocouple is immersed in the 2900°F bath. This requirement posed interesting problems in regard to thermal conductivity to the thermocouple tip and thermal insulation for the noble metal–base metal junction. The combination needs to act like a complete noble-metal thermocouple for just the few seconds it takes to make the measurement.

4. The tiny thermocouple tip is protected from contamination and physical damage as it is plunged through the furnace atmosphere and the surface slag.

5. The thermocouple holder, or lance, is light enough to be handled easily by one man and yet heavy enough to pierce the slag and metal surfaces and withstand general rough steelmill use.

Extensive development and design work, tooling, and testing ultimately produced the present expendable devices that sell for considerably less than $1 each. In the 13 years since their introduction, over 100 million units have been used in steelmills throughout the world. This was a case, then, of the basic foundation being available from classical physics; the challenging applications problems awaited imaginative, innovative design work before they were resolved.

As a second example of the contributions of basic physics to instrumentation, nuclear magnetic resonance (NMR) provides a good illustration of sophisticated basic research that resulted in the development of instrumentation that, through its use in both further research and field measurement applications, has been of major assistance in areas of substantial social benefit. It has benefited the U.S. balance of trade by providing exports not only of NMR instrumentation but also of new products particularly in areas related to chemistry. It has found increasingly important applications in biology and medicine in uses as diverse as understanding biological structures at the molecular level and the routine search for drug addiction evidence in urinalyses. The invention, development of the technique, and its commercialization have been a cooperative endeavor among the scientists—in many cases physicists—the applications scientists, the business administrators, and the production workers.

The phenomenon of NMR was discovered by Bloch at Stanford and Purcell at Harvard during the winter of 1945–1946 and earned for them the joint award of the 1952 Nobel Prize for Physics. The interest of these physicists was the determination of magnetic moments and angular momenta of various nuclear isotopes. The measurement is performed by placing the atoms of interest in an intense magnetic field and exciting and detecting absorption of energy as a function of its frequency. By measuring the frequency at which energy absorption takes place and relating it to the strength of the magnetic field, it becomes possible to determine the ratio of magnetic moment to angular momentum of the particular nuclear species in question. Since this information provides a characteristic signature not only for the nucleus involved but also, as was soon discovered, for the particular atomic or molecular environment in which it finds itself, the basis for a highly useful analytical tool was at hand.

It was soon found, too, that the measurement could be used in reverse to determine the strength of a magnetic field by relating it to the absorption frequency of a nucleus for which the gyromatic ratio was already known. Thus, as a significant by-product, a very accurate magnetometer was born.

The greatest impact of the NMR technique was found in chemistry. It came about because of the discovery, again by a group of physicists, that the exact gyromagnetic ratio of a given atom depended to a small extent on the chemical molecule in which the atom was located. Should several like atoms exist in the same molecule, several slightly different resonance frequencies are found. This finding was designated the chemical shift because its value is determined by the chemical environment of the atom. Studies of proton NMR spectra have been used extensively since 1952–1953 to determine the structure of molecules and the composition of mixtures and to understand equilibrium reactions. As instrumentation was improved by engineers and scientists, the technique was used for a wide variety of elements, for example, fluorine, phosphorus, boron, and nitrogen, in addition to the original hydrogen, and is now one of the most valuable instruments for chemical research. In fact, after a chemist synthesizes a new product, he can determine within hours many of its structural characteristics. In the latest development with superconducting magnets, Fourier transform techniques, and signal averaging devices, the method is being used in the biochemical field to study the structure of very complex proteins for the identification of various hydrogenous and other components of the complex molecule. The present state of development indicates that work with ^{13}C will make it possible to identify the different

[295]

carbon atom complexes and locations in complex organic molecules, again providing basic structural information for the research chemist and bio-chemist. The technique is now accepted as a standard method of chemical analysis and is used throughout the chemical industry providing a significant saving in time and cost.

Both NMR and its companion, electron spin resonance (ESR), discovered at about the same time by groups at Kazan and Oxford, are used to study biomolecules from simple amino acids and sugars through hormones, enzymes, and DNA. It is also used to study molecular conformation and structure, as well as metabolic reaction rates and mechanisms. A recent widely publicized application of NMR and ESR is for detection of a metabolic product of heroin that is found in the urine of a user, thus giving the first rapid and reliable screening test for drug addiction. This application marks only the beginning of widespread use of these techniques for similar purposes.

Since about 1955, NMR magnetometers have been used extensively to make airborne and ground-station measurements of the earth's magnetic field. These measurements are particularly useful to petroleum geologists in helping to identify formations in which oil might be found. Some of the early earth satellites carried proton magnetometers to determine the magnetic-field strength at large distances from the earth and relate it to the various theories concerning the origin of the earth's magnetic field. Several companies produce laboratory magnetometers for determining the precise field strength of laboratory magnets.

Fundamental to the application of NMR has been the conversion of a complex and often temperamental laboratory instrument, which required tender loving care from the physicists and skilled technicians who used it, to a reliable, rugged, commercially available instrument that when turned on could be expected to work consistently without delicate adjustment or special care. Here again the skills of the applications physicist and engineer have been brought to bear to provide an enormously useful new device.

Estimates of the time needed for the cost savings to pay for an NMR instrument range from a few months to a few years. The usefulness to industrial laboratories has been further demonstrated by the fact that practically all major chemical companies have multiple NMR instruments, some of them being multiple copies of a given type and others having differing characteristics.

The NMR instrument market for chemistry has grown rather steadily since the first sales in 1953. The market size in 1971 is estimated at $25 million, with approximately 40 percent of this market within the United States. Prior to 1966, almost the entire market was supplied by one U.S. company. At the present time, major suppliers of this market are found in Germany, England, and Japan, as well as in the United States. The ratio

[296]

of U.S. exports to imports of NMR equipment has dropped from approximately 10 in 1966 to approximately 1.5 at the present time.

Instrumentation Costs

Estimates of the escalation in the intrinsic costs of doing scientific research range from some 3.5 percent to 7.5 percent per year depending on the exact assumptions made. This figure is quite apart from any consideration of overall inflation or of salary escalation in the research enterprise; it relates only to the increased costs that reflect increased sophistication of the questions addressed and the measurements made. It has long been traditional wisdom in the physics community—and there are spectacular examples to support this belief—that much of this increase stemmed from the rapidly increasing cost of the necessary instrumentation. The Panel on Instrumentation examined this question and concluded that no such simple conclusion is warranted.

Well-Established Instruments   Instruments that were already well established a few years ago, such as certain types of test instruments and recorders, successive versions of which simply reflect improvements in design details, have had inflationary price increases during the past ten years that range from 25 percent to 50 percent. Certainly for such instruments the impression of higher prices would be confirmed.

Innovative Instruments   Instruments that a decade ago were relatively new in the marketplace and reflected major steps forward in complex instrumentation, such as NMR analyzers, have decreased steadily in price in recent years—for the same performance capability—moving significantly counter to inflationary trends. A basic assembly that sold for, say, $50,000 a decade ago might sell today for 60 percent of that amount. This pricing trend is in contradiction to the general impression of price increases. However, a research scientist might not be satisfied to purchase currently just such a basic instrument; rather, he might desire—or need—today's most advanced design to take advantage of its greater capabilities. Such an advanced unit might well sell at the same price or moderately higher than did the only unit available some ten years ago. In addition, it might be equipped with additional accessories, at a still higher price.

Computers   Computers and their constituent elements are considerably less expensive today than they were a decade ago. Figure 4.111 illustrates the drop in price in core memories from about $0.07 per bit in 1966 to less

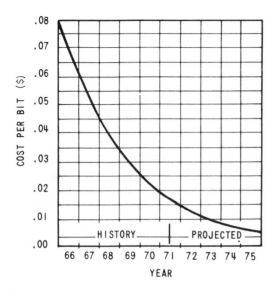

FIGURE 4.111   Core memory system cost.

than $0.02 per bit in 1971. Similarly the decrease in the price in mini-computers from about $27,000 each in 1963 to the $5000–$7000 range in 1971 is shown in Figure 4.112.

In each case, these decreases reflect advances in the technology of design and fabrication and substantial increases in production volume. Thus if a research scientist were to purchase a bare computer today he would be spending much less for it than he would have spent a few years ago. This, again, apparently contradicts the impression of increased prices for instrumentation. However, the current substantially lower price for a computer frequently encourages its purchase by a scientist who might not have considered such an acquisition when their cost was greater.

Advanced Systems Combinations   It is possible that advanced systems combinations have been most responsible for the impression of high current equipment costs. Automated system assemblies, incorporating computer direction and computer data processing, have been developed, with capabilities far beyond those available even a few years ago.

It is generally true that an individual researcher spends more for his instrumentation—for outfitting his individual laboratory—than was the case even a few years ago; but he may be getting a great deal more than he could have acquired at a lower cost a few years ago. In addition, he may be achieving secondary savings due to conservation of his own time and possibly the time of technician assistants through the automation of

[298]

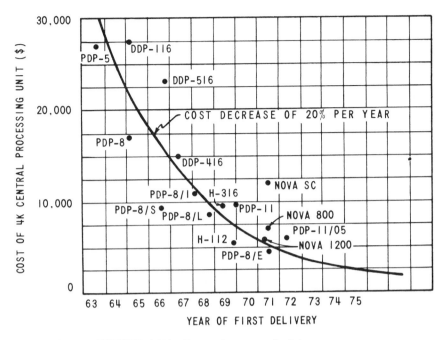

FIGURE 4.112   Decreasing costs of minicomputers.

computational and analysis steps. In terms of the cost per unit of new information, however inadequately this may be defined, there appears to be little question that a significant reduction has occurred in recent years. However, this reduction has been coupled with the need for more information to establish each new scientific finding as man probes ever deeper into nature. Despite the best efforts of the instrumentation industry to provide more capability for unit cost in its products, the cost of acquiring new information increases at the rates indicated earlier in this section.

Problems in U.S. Instrumentation Activities

As indicated above, U.S. instrumentation, at least since the mid-1940's, has enjoyed a position of international pre-eminence. Progress has been exceedingly rapid, spurred by demands from a burgeoning technological society and a vigorous scientific and engineering community. Both progress and pre-eminence are currently in danger.

The change in the growth rate of support for physics has had at least four important effects on instrumentation activity in the United States:

[299]

1. The scaling-down of research effort suggests that fewer new concepts and ideas fundamental to new instrumentation will become available in any given period; it also implies decreased pressure on the instrumentation industry for new instruments.

2. Although the instrumentation industry has a strong record of in-house research and development, the instrumentation groups in national laboratories and some of the larger universities have served traditionally as major sources of new ideas and devices that were then picked up, modified, and marketed by the appropriate industry. Under conditions of limited support these instrumentation groups, assembled patiently over many years, have been early casualties and this source of instrumentation innovation essentially terminated. Unless means are developed to reinstate such activities it will be difficult indeed to maintain the United States in a competitive position in the face of burgeoning activity in Europe and Japan.

3. With limited funding there is a natural tendency in any laboratory to live on instrumentation capital—to defer the purchase of a new instrument, however badly needed, in preference to discharging yet another research staff member. In the long term, this practice can have serious ill effects on the research activity; in the short term it can have a devastating effect, particularly on smaller, highly specialized instrumentation industries whose markets effectively evaporate. Many have been forced into bankruptcy in the past few years. Unless current trends are reversed, more such firms probably will face bankruptcy. These industries represent an important national resource and one whose loss cannot be viewed lightly.

4. In the past decade, there has been a discernable de-emphasis in the education and training of instrumentation scientists and engineers in U.S. universities and colleges. It is in just this activity of exploiting scientific phenomena in new engineering and technological areas that the United States has been enormously successful in the past; it is in this type of activity that this nation must continue to remain at the forefront if it is to preserve not only an internationally competitive economic posture but also the capability to respond to the challenge of improving the environment and the quality of life. There is a serious need for re-examination of U.S. education programs related to instrumentation development and application.

> Here, too, we may again repeat what we have said
> above concerning the extending of natural philosophy
> —so as to prevent any schism or dismembering of
> the sciences; without which we cannot hope to
> advance.
>
> FRANCIS BACON (1561–1626)
> *Novum Organum*, Book I, 107

THE UNITY OF PHYSICS

Introduction

This chapter has been concerned thus far with the individual subfields of physics, partly for convenience, partly to bring out the internal logic in each. To stop at this point, however, would be to neglect perhaps the most important aspect of physics—and one all too frequently forgotten: The study of natural phenomena proceeds simultaneously on many frontiers each of which supports and nourishes the other. There is a very broad spectrum of research activities in physics ranging from the most basic to the most applied—from almost exclusively intrinsic to equally exclusively extrinsic activities with all intervening gradations. This diversity, amid overall continuity, is the hallmark of a vigorous science and has long been recognized as such. Its application to physics is simply typical of its more general truth throughout all of science.* Preservation of this unity in science and in physics is of vital importance and is an essential consideration in any national planning for science.

With increasing specialization what tends to be forgotten is the remarkable extent to which even the most intrinsic techniques and concepts diffuse throughout physics and the remarkable degree of commonality that exists. The frequently discussed fragmentation of physics has been much exaggerated.

In the nineteenth century, the unifying concepts were those of Newtonian mechanics; in the twentieth century, the two unifying principles have been relativity and quantum mechanics. In fact, the latter has played by far the dominant role for the simple reason that, in the usual terrestrial physical phenomena, relativistic considerations appear as fundamental but nonetheless small corrections and perturbations; elementary-particle physics is clearly an exception. Increasingly throughout physics, and more recently in both chemistry and biology, quantum mechanics has provided the overall unifying concept.

To illustrate this unity we draw on data from the physics of condensed matter, atomic physics, nuclear physics, and elementary-particle physics.

---

* See Chapter 3.

The degree to which the latter three subfields show parallel spectroscopies was first emphasized by Weisskopf.* Here his discussion is expanded to include the physics of condensed matter as well.

The Four Spectroscopies

Figure 4.113 is a highly schematic illustration of the systems considered in this section. In condensed matter the exciton is selected as the elementary excitation. The exciton is simply the composite entity resulting

* See, for example, V. Weisskopf, "The Three Spectroscopies," Scientific American, *218*, 15 (May 1968).

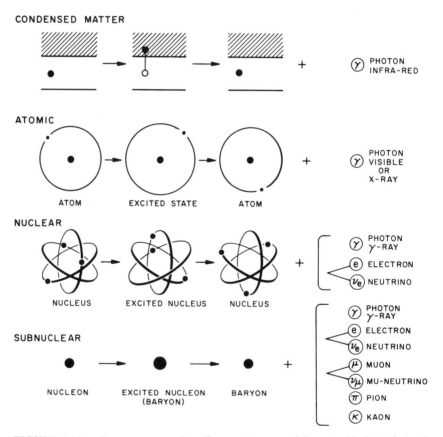

FIGURE 4.113   Four spectroscopies. [Source (for part of figure): V. A. Weisskopf, "The Three Spectroscopies," Scientific American, *218*, 16 (May 1968).]

from the interaction of an electron, in the conduction band of a semiconductor, bound to a hole in the band gap. When the electron and hole recombine, the energy involved is radiated as an infrared photon. In the condensed-matter and atomic examples, the energies involved are not sufficiently high to permit radiation of other than photons of appropriate wavelength. In the nuclear example, once the available energy exceeds 1.022 MeV (the rest mass of an electron–positron pair) it becomes possible to radiate such a pair instead of a photon; in certain situations in which angular momentum selection rules forbid photon emission (as from a O^+ to a O^+ state), pair emission is the only mode of de-excitation available to the nuclear system. Such decays clearly occur via the electromagnetic interaction.

In the nuclear system, the weak interaction—typically weaker by a factor of roughly 10^9, and one of the two new natural interactions first encountered in nuclear physics—can also act to de-excite nuclear states. Wherever it is in direct competition with the electromagnetic interaction, the electromagnetic decay modes are completely dominant. Under certain conditions, however, selection rules strongly hinder the electromagnetic interactions, and the weak-interaction decay modes, in which an electron and an antineutrino (or a positron and a neutrino) are emitted, become dominant.

In moving to the much higher energies characteristic of elementary-particle physics, clearly these electromagnetic and weak-interaction decay modes remain; however, as soon as the available energy becomes greater than the rest mass of the muon, the weak-interaction mode involving emission of the muon (heavy electron) and its corresponding mu-neutrino becomes possible. In going to larger available energies, as soon as the rest mass of the pion is exceeded its emission becomes possible; because pion emission takes place via the strong nuclear interaction (typically 100 times stronger than the electromagnetic interaction, and the second new natural force encountered in nuclear physics), it is a dominant decay mode unless, again, selection rules inhibit or preclude it. With still higher available energies, of course, it becomes possible, through the strong interaction, to emit the heavier kaons and other mesons as well. It may well be that with increasing energy an even heavier electron than the muon will be discovered under the mantle of the weak interaction. What is evident from this description is the systematic opening of new decay modes as ever more energy becomes available.

Next we illustrate the striking similarity in the actual spectroscopy of the different subfields. Figure 4.114 compares the atomic and nuclear spectra of sodium. Plotted here is simply the excitation energy of the quantum states of the atomic and nuclear systems—in essence the energies

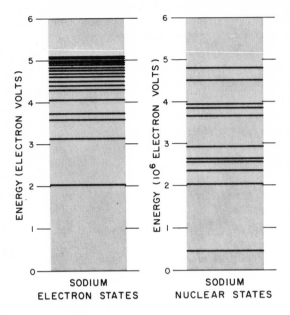

FIGURE 4.114 Atomic and nuclear spectra of sodium are similar in character. But the atomic spectrum (left) can be plotted on a scale whose units are electron volts, whereas the spectrum of nuclear states (right) requires a scale whose units are larger by a factor of 10^6. [Source: V. A. Weisskopf, "The Three Spectroscopies," Scientific American, *218*, 21 (May 1968).]

corresponding to the different solutions to the quantum-mechanical Schrödinger equation for each of these two systems.

What must be emphasized is that knowledge of these solutions comes from experimental observations, not from actual solution of such equations. The reason is obvious but perhaps bears repeating. Even assuming that physicists had complete knowledge of the forces involved—as, indeed, they do in regard to the atom, where the force is purely electromagnetic, but do not in regard to the nucleus, where as yet there is only approximate knowledge of the strong nuclear forces involved—the necessary mathematics for the exact formulation of the many-body problems involved does not exist. Of course, one can write approximate equations and, with the help of ever larger computers, obtain solutions of ever-increasing precision, granting the validity of the assumed forces. But in the case of the atom, because it is vastly simpler and less expensive, and in the case of the nucleus, because lacking better knowledge of the forces involved there is no option, physicists locate the quantum states, the solutions of implicit Schrödinger equations, experimentally in that branch of each subfield labeled spectroscopy. With the sophisticated techniques now available, it is possible not only to locate these solutions in energy (determination of the eigenvalues) but also to determine each of the quantum numbers labeling the individual solutions and, to an increasing extent, from a wide variety of measurements, to obtain increasingly detailed information on the structure of the state (determination of the eigenfunctions).

The spectra of Figure 4.114 illustrate, first, the remarkable similarity between the two excitation spectra, with an ordinate scale change of 10^6. They also illustrate a characteristic difference between spectra for the long-range atomic and the short-range nuclear forces: in the former there is a sharp break between the bound states and those in the continuum, whereas no such break occurs in the latter. Parts (a) through (d) of Figure 4.115 are corresponding presentations of the excitation spectra for exitons in cadmium selenide, a potassium atom, an aluminum nucleus (^{25}Al), and the baryon family of elementary particles (see also the Report of the Panel on Elementary-Particle Physics in Volume II for the corresponding spectra for the meson family). In each case, the spectrum has been decoupled horizontally to display more clearly some of the internal grouping of states according to certain of the dominant quantum numbers. The similarity, apart from the ordinate change from millielectron volts to electron volts to millions of electron volts and finally to billions of electron volts, is again striking and serves as an excellent illustration of the common approach used throughout wide areas of physics. Fundamental understanding comes as the physical models, which in essence are nothing more than mathematically tractable approximations to the unknown Schrödinger equations, are adjusted and modified—and occasionally scrapped—to reproduce the spectroscopic data. A model that accomplished only this task would be of little value; its value is based on the extent to which it permits new insights, suggests new studies, and successfully predicts the results of measurements yet unmade.

As might be expected from Figure 4.115, to the extent that all the emitted radiations in the different systems are the same, the experimental as well as the theoretical techniques have much in common throughout all of physics; increasing energy, of course, poses its own special detection problems.

Thus far we have emphasized the commonality of approaches and techniques that unify physics; we now turn to a different aspect of this unity, the way in which the results and insights of many branches of physics are required for fundamental understanding of a given physical situation or object. We have chosen to illustrate this characteristic with reference to one of the most exciting new objects found in nature in the recent past—the pulsar—because it has attracted the interest and activity of a very wide range of leading physicists and because it draws on so

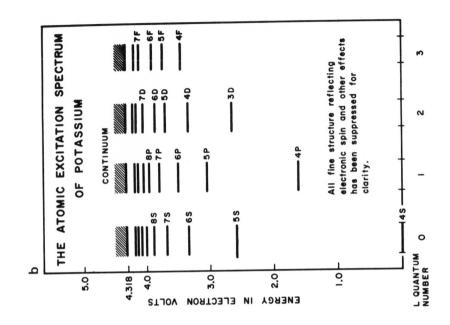

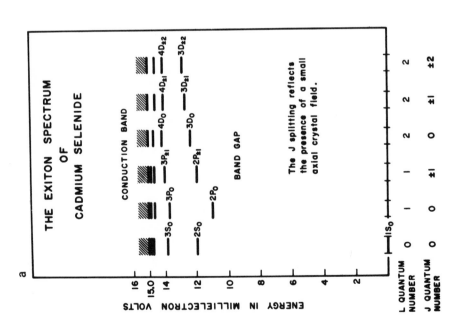

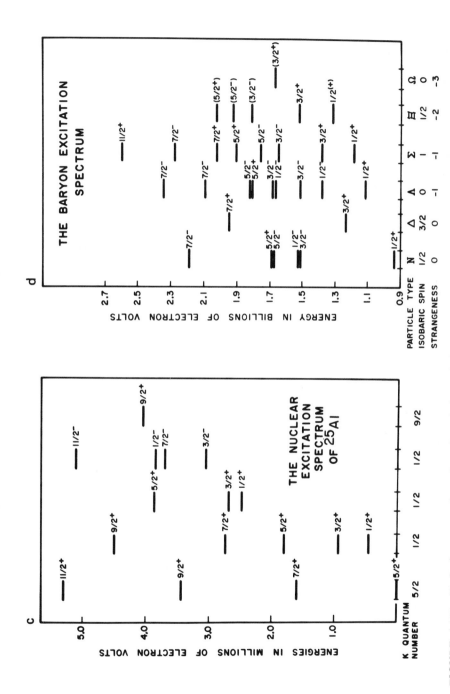

FIGURE 4.115 Excitation spectra for exitons in cadmium selenide, a potassium atom, an aluminium nucleus, and the baryon family of elementary particles.

many physical subfields for understanding of its structure and behavior. In considering this example we have drawn heavily on a recent article by Ruderman.*

The Pulsar

Perhaps the most exotic and varied forms of matter in the universe are to be found within the cores of dying stars. As a star evolves, its center grows hotter and denser. When it has finished burning its nuclear fuel, which supplied almost all of its radiated light energy, the star rapidly approaches its final state. Theory predicts three possibilities:

1. A star whose mass is less than about 1.4 times that of the sun can die as a familiar white dwarf. These are extremely dense stars about the size of the earth. Their central densities may exceed millions of grams per cubic centimeter. In such stars the enormous pull of gravity, which tends to crush the star, is balanced by the pressure from very rapidly moving electrons whose velocities may be comparable to that of light. Such high velocities are a direct consequence of the high density that forces electrons to be much more closely packed together than they are in the atoms of normal matter. The Pauli Exclusion Principle of quantum mechanics forbids identical electrons from getting close to each other unless their relative velocity is correspondingly large. This motion is the same as that which gives rise to the pressure that makes common solids difficult to compress and prevents the collapse not only of white dwarfs but of almost all forms of terrestrial matter. The different stages in a stellar collapse to and beyond the white dwarf phase are illustrated schematically in Plate 4. XVII.

2. A dying star may appear to contract forever toward a radius of a few kilometers and a density exceeding 10^{17} g/cc. This strange fate is predicted by the General Theory of Relativity for all stars more than twice as massive as our sun. Such relics are called black holes. Although they can attract to themselves whatever matter or radiation approaches them, the gravitational pull on their contracting surfaces has become so huge that not even light can escape. Black holes are probably common in the galaxy and throughout the universe, but they are hidden because of the great difficulties in observing them. The evolution of a black hole is shown schematically in Plate 4. XVIII.

3. Finally, the collapsing system may stabilize as a neutron star— a mass of a million earths compressed into a sphere barely capable of

* M. A. Ruderman, "Solid Stars," Scientific American, 224, 24 (Feb. 1971).

Matter's Struggle for Space in the Crush of Collapse

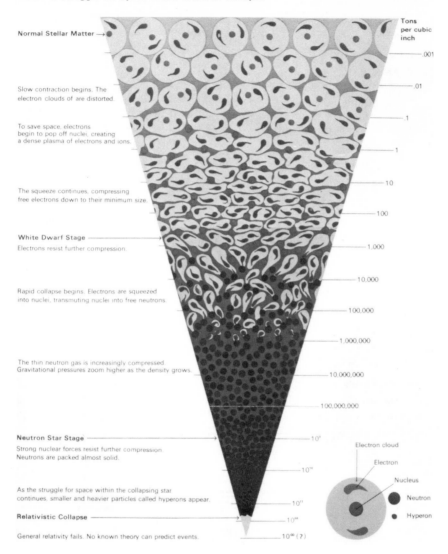

Normal Stellar Matter →

Slow contraction begins. The
electron clouds of are distorted.

To save space, electrons
begin to pop off nuclei, creating
a dense plasma of electrons and ions.

The squeeze continues, compressing
free electrons down to their minimum size.

White Dwarf Stage
Electrons resist further compression.

Rapid collapse begins. Electrons are squeezed
into nuclei, transmuting nuclei into free neutrons.

The thin neutron gas is increasingly compressed.
Gravitational pressures zoom higher as the density grows.

Neutron Star Stage
Strong nuclear forces resist further compression.
Neutrons are packed almost solid.

As the struggle for space within the collapsing star
continues, smaller and heavier particles called hyperons appear.

Relativistic Collapse

General relativity fails. No known theory can predict events.

Tons
per cubic
inch

.001
.01
.1
1
10
100
1,000
10,000
100,000
1,000,000
10,000,000
100,000,000
10^9
10^{10}
10^{11}
10^{12}
10^∞ (?)

Electron cloud
Electron
Nucleus
Neutron
Hyperon

PLATE 4.XVII Matter's struggle for space in the crush of collapse. The heavier a
particle, the smaller the volume it is able to occupy. Atoms in a shrinking star are
thus crushed, in turn, to electrons and nuclei, neutrons, and increasingly heavy
hyperons. Whether each can counter gravity depends on the star's mass: less than
1.4 suns, free electrons stop the slow compression; less than 2 suns, neutrons will halt
the collapse; more than 2 suns, no particle is able to resist gravity. [Source: *Science
Year. The World Book Science Annual.* Copyright © 1968, Field Enterprises Educational Corporation.]

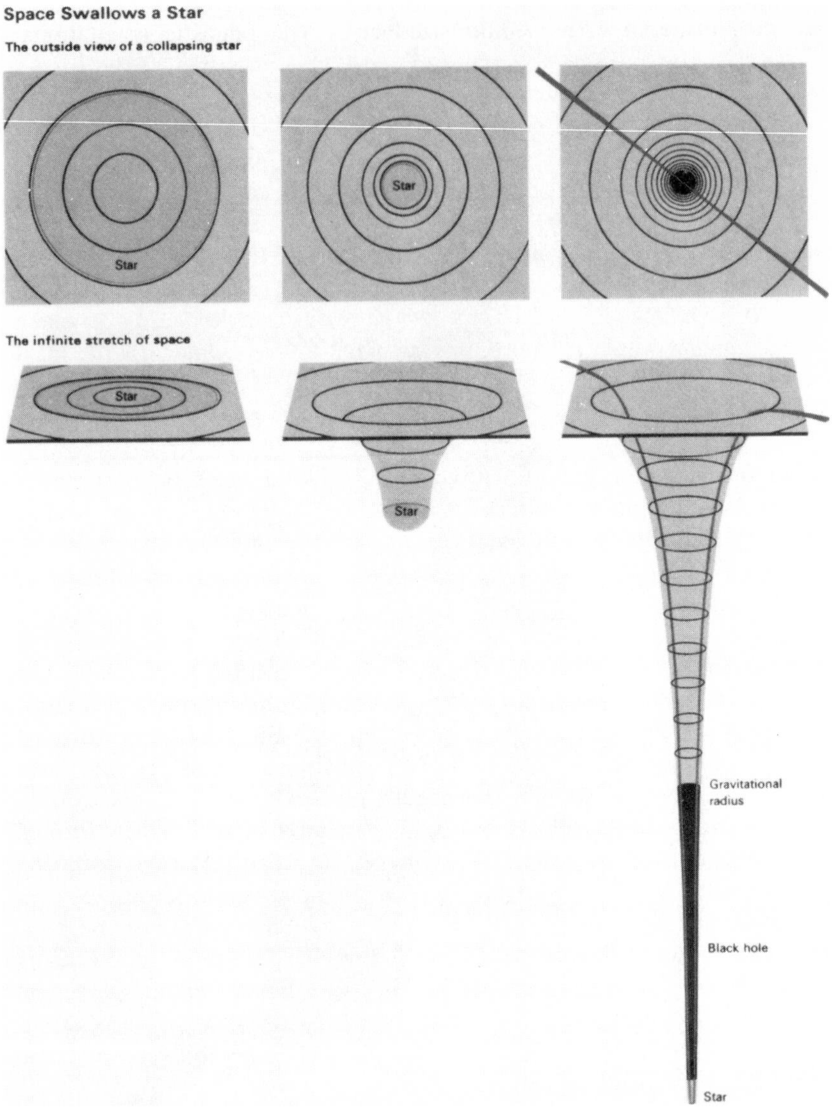

Space Swallows a Star

The outside view of a collapsing star

The infinite stretch of space

Gravitational radius

Black hole

Star

PLATE 4.XVIII Space swallows a star. Distances near a collapsing star are stretched by the star's rising density. To an outside observer, the star merely shrinks (top row). The originally flat surface of this simplified two-dimensional space is rapidly stretching, however (bottom row), increasing the distance to the star. Thus, the line through the right drawing is actually much longer than it seems from the outside. In three-dimensional space, the star would be shown within a cube, instead of a square, and as the star collapses, the distance through the center of the cube would stretch to infinity. [Source: *Science Year. The World Book Science Annual.* Copyright © 1968, Field Enterprises Educational Corporation.]

[310]

containing the area of a medium-sized city. The fantastic gravitational attraction at the surface of a neutron star (10^{11} times that at the earth's surface) is balanced by the same combination of nucleon motion and repulsive forces that keep atomic nuclei from collapsing despite the powerful attractive nuclear forces that hold them together. Since matter must be squeezed until different nuclei touch each other before such repulsive forces become effective in preventing stellar collapse, a neutron star must have a central density near to or greater than the matter within such nuclei. Therefore, the core density of a neutron star exceeds 10^{11} g/cc (or $\sim 10^8$ tons per cubic inch), an almost inconceivable density at which a speck of sand would be more massive than an ocean liner. If the mass of a neutron star is less than one sixth that of the sun, the central density is too low to support it stably against the pull of gravity, thus it would pop out to remain a white dwarf. If, on the other hand, it is considerably heavier than the sun, gravity will crush it toward a black hole no matter how repulsive the nuclear forces are at a short range. Here is the strange case in which the gravitational forces, intrinsically weaker than the nuclear forces by the enormous factor of 10^{40}, still dominate because of the great concentration of matter.

The rapidly growing body of evidence that suggests identification of pulsars as rapidly rotating neutron stars gives a clue to their abundance and genesis. The association of young pulsars with supernova remnants, as in the now familiar example of the Crab (see Figure 4.116), suggests that neutron stars are formed in violent supernova explosions that have already consumed about one thousandth of all the stars in our galaxy. Many or most such explosions result in neutron stars. When formed they are extraordinarily hot, well above 10^{11} deg. Various neutrino emission processes should quickly cool them so that within a few centuries their internal temperatures would drop to a few hundred million degrees. Remarkably, the behavior of this very hot stellar interior is in many respects exactly analogous to that of terrestrial liquid helium when it is cooled close to absolute zero; in terms of the phenomena expected within a neutron star interior—superconductivity and superfluidity—it is the coldest known place in the universe.

Constituents of a Neutron Star

A 10-km traverse from the surface to the center of a neutron star would take a traveler through all densities from a near vacuum to well above 10^{11} g/cc. The corresponding pressures would extend from zero to above 10^{28} atm. Present knowledge of how the pressure and constituents of

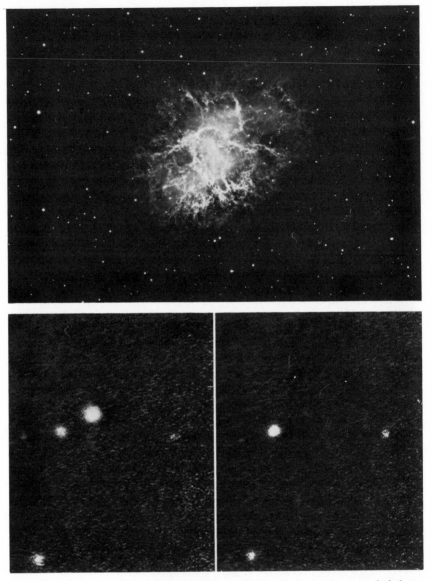

FIGURE 4.116 The Crab nebula. Within the Crab nebula (top), an exploded star is a pulsar that emits visible light—30 bursts a second. It was thought to be a normal star, but high-speed electronic imaging (below, left) shows that the source is off more than 97 percent of the time (below right). [Source: Photograph (top) courtesy of the Hale Observatories. Photographs (bottom) from: *Science Year. The World Book Science Annual*. Copyright © 1969, Field Enterprises Educational Corporation.]

matter vary with increasing density is represented in Figures 4.117 and 4.118. The different portions of the pressure versus density equation-of-state curve are contributed by physicists in varied specialties. From A (matter at negligible pressure) to B (about the pressure at the center of the moon), the equation of state can be measured in the laboratory and, of course, depends sensitively on the chemical nature of the material. From B to C, a combination of a mathematical model, evolved from one proposed about 40 years ago by E. Fermi and L. H. Thomas, together with the use of modern high-speed computers, yields an adequate theoretical description. Beyond the point C, matter is so compressed that the Pauli Exclusion Principle compels the electrons to have too much kinetic energy to remain bound to their nuclei as in the case of normal matter. These high-speed free electrons then contribute almost all of the pressure; the equation-of-state curve can be calculated accurately using modern techniques of many-body quantum theory. Near point D, the electron velocities approach that of light so that a correct description of such matter must utilize the Dirac relativistic quantum-mechanical equations. At a density of about 10^9 g/cc, the electron energy becomes so large that the tightly bound protons within the atomic nuclei absorb some of the more energetic electrons. Such protons are thus converted to neutrons (plus neutrinos, which easily escape) in a process that is the precise reverse of the normal process wherein a free neutron decays with a

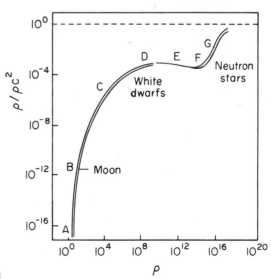

FIGURE 4.117 The dimensionless ratio of pressure (p) to density (ρ) times the speed of light squared (c^2) as a function of density, in grams per cubic centimeter. (The doubled portion of the curve corresponds to matter stiff enough to keep a star from collapsing under its own weight. Neutron stars have central cores whose pressure and density lie above the point F. White dwarfs, planets, etc., have corse below D.) [Courtesy of M. A. Ruderman.]

[313]

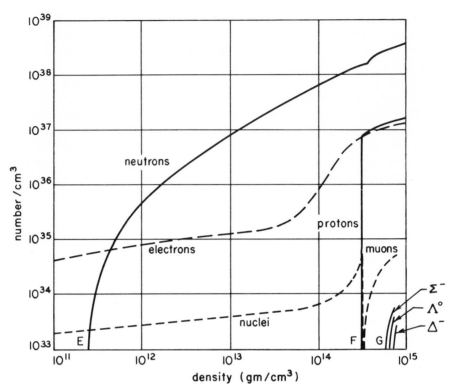

FIGURE 4.118   The constituents of matter as a function of density. The letters, E, F, G, refer to points on Figure 4.116 and are described in the text. Σ^-, Λ^0, Δ^- are the names of heavy unstable nucleons. The muon is a very heavy unstable electron. [Source: M. A. Ruderman, "Solid Stars," Scientific American, 224, 24 (February 1971).]

lifetime of 12 min, to leave a proton and an electron antineutrino pair, and the nuclei in such superdense matter are much more neutron-rich than are stable nuclei in normal matter. As matter is squeezed to still higher densities, the electrons continue to be absorbed by protons, converting them to neutrons. The pressure from the rapidly moving electrons no longer rises strongly with increasing density because of their disappearance into the neutrons, so that in this regime matter is relatively more compressible than it is at lower densities. At the point E, corresponding to a density of 3×10^{11} g/cc, the nuclei are so neutron-rich that they "drip" neutrons into the interstitial volume between the nuclei. Beyond E, matter consists of free electrons and nuclei embedded in a sea of neutrons. These nuclei contain two or three times more neutrons than any isotopes found on earth. Continued compression causes further absorption of electrons by protons and their conversion into neutrons. At densities approaching 10^{11} g/cc, almost all of the matter density is composed of the free neutrons, which are about 25 times more abundant than the protons (all of which are bound into nuclei) or electrons. At 3×10^{11} g/cc (point F), the nuclei rather suddenly dissolve, and such ultradense matter consists mainly of a sea of neutrons interpenetrated by a much less dense sea of unbound protons and relativistic electrons. A further increase in density to point G results in the conversion of some of these particles to other kinds of elementary particles, which, on earth, are produced only in the largest particle accelerators. These particles, when isolated, are very unstable, with lifetimes between 10^{-6} and 10^{-22} sec. However, in the superdense environment within a neutron star, they can be stable because the Pauli Exclusion Principle prevents their decay products from being injected into an already existing sea of similar particles and the decays do not occur. The equation of state of this sort of matter is not yet known. For denser stars most of the total stellar mass may consist of these normally unstable strange particles. During the collapse, as illustrated also in Figure 4.117, the stellar matter passes from the atomic regime of condensed matter and atomic physics through the nuclear realm, where, incidentally, nuclear matter—long a favorite idealized concept of nuclear physicists—emerges in kilometer-scale natural samples, to the still mysterious realms of elementary particles —some already known and some yet to be discovered in the march to ever higher energies. In this one collapse, the star sweeps through much of modern physics.

The original name, neutron star, is generally used for these collapsed objects, even though other elementary particles may well be present, because a typical collapsed star would be expected to contain mainly neutrons, especially in its deep interior. The nuclear force interactions between pairs of neutrons are exactly analogous to those between pairs of electrons in a normal low-temperature solid that has become a super-conductor, except that they are much stronger, thus causing such a transition at enormously higher temperatures (Figure 4.119). Because the neutrons are uncharged, the neutron sea is expected to become only a superfluid, without spectacular electrical properties. It can conduct heat fantastically quickly, and it can flow as if without any viscosity. Normal liquid helium, when cooled to within a few degrees of absolute zero, has such properties. But the closest terrestrial analogue is a fluid composed of the helium isotope ${}^3\text{He}$. However, it has not yet been possible to cool it to a low enough temperature (much less than a thousandth of a degree above

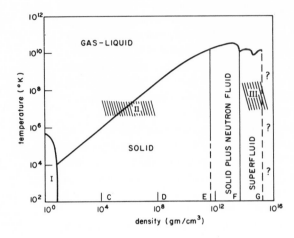

FIGURE 4.119 Phases of superdense matter at various densities and temperatures. The nuclei are assumed to be the most stable ones for a given environment. (At densities above 3×10^{11} g/cc this would occur quickly even at zero temperature.) Region I is, roughly, that accessible in the laboratory. Region III corresponds to neutron star cores. Region II is that for the centers of white dwarfs. The point C is that for the center of our sun, which will ultimately move up and toward the right. [Source: M. A. Ruderman, "Solid Stars," Scientific American, *224*, 24 (February 1971).]

absolute zero) to mimic the behavior expected in the neutron fluid of a neutron star at temperatures of a few hundred million degrees. The free charged protons in the deep interior of neutron stars are expected to form a superconductor. Currents once set up maintain themselves forever; a magnetic field trapped within such matter will never die out. But even if the interior protons were not superconducting, the electrical conductivity contributed by the relativistic electrons in the crust and core is over a million times greater than that of copper. This characteristic alone is sufficient to hold an original magnetic field for over tens of millions of years. The entire physics of condensed matter contributes to the understanding of these phenomena.

At densities approaching 10^{15} g/cc, expected in the central core of heavier neutron stars, neither the composition nor the structure is yet fully understood. Both depend on a deeper knowledge of the properties of elementary particles and their interactions than is presently available. Here, if anywhere, is a region that might contain surprises.

Neutron stars are probably spinning very rapidly when formed, perhaps up to a thousand revolutions per second. Fast rotation rates are inferred from observations of present pulsar periods and their rate of increase, as well as from the conservation of angular momentum in a rapidly collapsing star. An interior neutron superfluid cannot rotate like a normal fluid. Rather it must set up an array of tiny vortices (whirlpools) whose rotational axes tend to be parallel to the rotational axis of the fluid as a whole. In a neutron star rotating about once a second—characteristic of pulsars—but with no external torques tending to slow its spin, the interior neutron superfluid would set up about 10^1 vortices per square centimeter in a triangular array. Averaged over many vortices, the motion of the fluid is essentially that of a rotating rigid body. But on a microscopic scale, the motion entails an enormous array of tiny whirlpools whose centers corotate with the star. Here many of the phenomena of fluid dynamics occur.

The laboratory analogue of the hot rotating neutron star is a cold hollow steel shell, with a thickness of about one tenth its radius, filled with low-temperature corotating liquid helium. At the center there may be a core with unknown and perhaps unexpected properties. The crust should be approximately $10°$ above absolute zero and the superfluid only a few hundredths of a degree. However, this rather remarkable model of a neutron star is complicated and enriched by an enormous magnetic field that threads the star and its immediate environment.

Structure

Nuclear physics and elementary-particle physics determine the composition of ultradense matter in neutron stars. But experiments and theories of condensed matter at temperatures close to absolute zero describe the organization and behavior of this matter (Figure 4.120). The outermost layer of a neutron star is a hot gaseous liquid whose density reaches 10^6 g/cc about a meter below the surface. Such matter is very much like that found throughout the interior of many conventional stars, especially white dwarfs. Below this region, and until a density of 3×10^{14} g/cc is reached, at which all nuclei have dissolved, is the solid crust in which the nuclei arrange themselves in a crystalline lattice much like those in terrestrial solids below their melting points. The computed melting temperature of superdense matter can exceed 10^{10} deg as long as nuclei remain to form a lattice. This enormous transition temperature is a consequence of the strong electrical (Coulomb) repulsion among nuclei that have lost their atomic electrons and are pushed very close together. Because the present temperature within a neutron star is far below the lattice melting temperature, the outer kilometer or so will form a solid crust. Relative to its melting temperature, the neutron star crust is much colder than that of the earth. It is up to 10^{18} times more rigid than a piece of steel and over 10^{20} times more incompressible. In the inner regions of the crust, where a neutron fluid fills the space between the nuclei, it is easy for the nuclei to change the number of neutrons they contain as well as their charge by electron emission or absorption. Therefore, any impurity

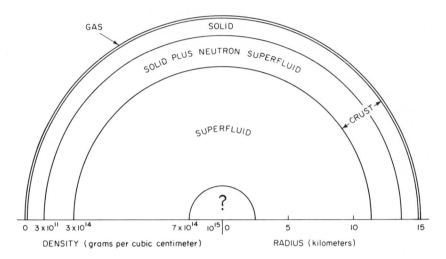

FIGURE 4.120 The structure of a neutron star. The thickness of the gaseous atmosphere is less than a thousandth of the stellar radius. The very lightest neutron stars, whose mass is near a sixth that of the sun, would probably have a crust that extends right to the center. Their total radii will also be much greater than that of the typical neutron star shown here. [Source: M. A. Ruderman, "Solid Stars," *Scientific American*, *224*, 24 (February 1971).]

nucleus whose charge or mass is inappropriate for the most stable nuclei at a given density will quickly change to one that is appropriate. In this way impurities disappear. A neutron star crust is thus the purest as well as the strongest matter known in the universe.

Dynamics

Even before pulsars were discovered it had been conjectured that, if a neutron star contained within itself the magnetic flux it had before its collapse, then it might possess a huge 10^{12}-G magnetic field. (The earth's field is less than a gauss.) The slow lengthening of pulsar periods suggests that neutron stars, at least those observed as pulsars, do have such enormous fields. Such magnetic fields can give sufficient radiation or coupling to the surrounding plasma to account quantitatively for the lost rotation energy. Ultimately this energy seems to appear in the production of very energetic cosmic rays; most of the cosmic rays that continuously bombard the earth may originate from such spinning neutron stars. The magnetic field profoundly affects the exterior, surface, and interior of a rotating neutron star. Outside the star it controls the coherent flow of

[318]

charged particles that gives the fantastically powerful radio emission that makes pulsars observable. Understanding of the behavior of a relativistic plasma in such a huge field is still very limited. At the stellar surface, the 10^{12}-G field completely changes the nature of whatever atoms are present. The stellar rotation coupled to the magnetic field can generate electric fields approaching an astonishing 10^{12} V/cc between the poles and the equator. Such electric fields will pull charged particles from the surface into a surrounding plasma. Finally, such an enormous magnetic field will strongly couple all electrically conducting regions within and outside the star. The plasma, crust, and core electrons and protons all corotate and slow together. It is their period of rotation that is measured in pulsars. But an interior neutron superfluid would probably rotate slightly faster, because it is generally difficult to slow a rotating superfluid. A slight jump in the crust angular rotation velocity would cause it to rub on the neutron superfluid, but so weakly that it might take weeks or years to return to its original spin rate. There seems to be support in observed pulsar periods for such behavior.

The theory of a slowing neutron star is still very primitive. There are many crucial unanswered questions: Exactly how does the spinning crust couple to the neutron fluid? How does an unknown central core react? Do the parallel vortex lines in the neutron superfluid tend to bend and twist to form a pseudo-turbulent interior superfluid? How does the crust respond to the shape changes forced on it by the varying centrifugal forces as the neutron-star spin slows down? Does the crust creep and flow plastically, or does it crack and give starquakes? How exotic can the behavior of the crust be? Could there be analogues of terrestrial volcanoes and mountain building?

Very small sudden changes in pulsar period equivalent to those that could be caused by surface motions of 10^{-3} cm have been detected. Such marvelously accurate observations, together with more refined theories of the various neutron star regions, will ultimately lead to a more precise description of spinning neutron stars. But enough is already known to show that the neutron star has a unique and wonderful structure that causes remarkable behavior, like that of no other object in our universe, and that calls on almost every branch of physics for its understanding.

Conclusion

The physicist who explores the nature of his immediate world learns directly only about a very rare and special part of the universe that can support life. Most of the rest is far too hot or cold or dense or empty. Knowledge of the structure and phenomena within most of the universe is based on an immense extrapolation from immediate surroundings to regimes of density and temperatures far removed not only from those found on earth but even enormously beyond those achievable in the laboratory. Such an extrapolation is sustained by experiments and theory embracing most subfields of physics. An understanding of many fantastic objects observed beyond the earth depends, in part, on the physics of normal condensed matter at temperatures near and below the lowest ever achieved on earth. Knowledge of the astrophysical universe uses nuclear physics in almost all of the regimes accessible in the laboratory; it also exploits high-energy physics and the theory of elementary particles at energies even beyond those reached in the largest accelerators. Plasma and fluid physics, as well as the surface physics of tiny dust particles, and the general theory of relativity are necessary contributors to the history of a star, the former to its birth as a condensation from a primeval gas, the latter to its death as an ultradense stellar cinder.

We shall return to this question of the unity of physics, and of science, repeatedly throughout this Report. The examples chosen for inclusion here are particularly striking; they are by no means isolated cases.

APPENDIX 4.A.   PROGRAM ELEMENTS

In the following tables, subfields of physics have been divided into program elements (major research areas). Although the definition of this term is somewhat imprecise, the intent has been to identify separable components of the subfields—components sufficiently large to have some internal coherence and reasonable boundaries and for which it might be possible to estimate present funding levels and the PhD manpower involved. As discussed in Chapter 5, the purpose of this exercise was to divide the subfields into units of activity that the Committee could rate in terms of intrinsic, extrinsic, and structural criteria. The purpose of the ratings was to test the feasibility of arriving at a consensus regarding the desirable relative emphasis among subfields and among program elements within each subfield.

Data available to the Committee permitted identification of program elements, at least roughly, for all subfields except earth and planetary physics. In some cases, the program elements cover the major activities in the subfield; in others, they do not. It should be noted that much of the work in such subfields as optics and acoustics lies largely outside physics, and that the basic physics research in some of the program elements involves only a small part of the total dollars and manpower associated with the subfield.

Clearly, most of the program elements could be further subdivided, but, to keep the overall number for the Committee's consideration within manageable limits, the number per subfield was somewhat arbitrarily restricted to approximately ten. As expected, the subfield panels found it convenient to make the divisions into program elements along different lines. For example, in elementary-particle physics the division is made in terms of the small number of major facilities and associated programs, whereas in condensed matter the division accents specific areas of research such as superconductivity. The program elements for astrophysics and relativity identify emerging areas of research that will require greatly increased funding. In the projected program for this subfield, the costs of satellites and large facilities are included. Funding figures associated with program elements in the other subfields do not include construction costs of major facilities.

In developing these program elements, the Committee worked with the panel chairmen; however, in some cases the elements used here are not identical with those suggested by the panel chairmen. In elementary-particle physics and nuclear physics it was possible to assign funding levels and manpower rather precisely. Similar assignments for some of the program elements in the other subfields may be in error by a factor of 2.

Elementary-Particle Physics (1971)

Program Elements	Federal Support [a] (Operations and Equipment) ($Millions)	PhD Manpower [b]
1. Accelerator developments	3.6	— [d]
2. National Accelerator Laboratory	— [c]	—
3. Other major facilities (e.g., AGS improvement project)	1.5	— [d]
4. Stanford Linear Accelerator	27	99
5. Brookhaven AGS	27	99
6. Argonne ZGS	20	65
7. Berkeley Bevatron	26	100
8. Cornell Synchrotron	3	20
9. CEA Bypass Storage Ring	2.3	11
10. University groups	37	1245 [e]

[a] Construction costs not included.
[b] Includes approximately 300 scientists having the following specialties or combinations of them: computer employment in research, accelerator design and development, accelerator operation, device design and development, and emulsion experiments.
[c] No NAL figures are given for FY 1971 since the accelerator will not be in operation until FY 1972. When in full operation, operating plus equipment funds required are estimated at about $60 million.
[d] Manpower included in elements 4 to 9.
[e] Includes approximately 245 PhD's doing particle research but not supported directly by federal funds.

Description of Program Elements

1. These activities are an integral part of the ongoing work at each of the major accelerator laboratories. They have a creative content quite apart from the particle research itself, although neither can progress without the other. The technological requirements lead to innovation and development in such fields as radio frequency, engineering, superconducting magnets, ultrafast electronics, computer technology, radiation-detection instruments, pattern recognition, and particle orbit theory.

2. 200–500-GeV proton accelerator to be for some years the only controlled source of protons in the world for research * in the energy range above 80 GeV and the only one in the United States above 33 GeV. Also includes in-house research activities comprising a small fraction of the particle research to be carried out at the accelerator.

3. These are major additions to the capabilities of accelerators—other than NAL— that have been planned or under construction for some years and are now complete or nearing completion. Includes: the major modification of the AGS to increase its intensity and other capabilities, the SPEAR storage ring at SLAC and the 12-ft liquid hydrogen bubble chamber at the ZGS. Each facility offers unique opportunities to open new areas of research but will require incremental operating and equipment funds for the purpose. Construction costs are not included.

4. 22-GeV electron accelerator, which is the only controlled source of electrons in the world for research * in the energy range above 10 GeV. Also includes in-house research activities comprising a substantial fraction (about one half) of the research carried out with this acclerator.

5. The 33-GeV proton accelerator at BNL, which is the principal source in the United States for research * using protons in energy range 12–33 GeV. Includes in-house research comprising about 25 percent of the total research activity.

6. The 12.5-GeV accelerator at ANL, which is the principal source in the United States for research * using protons in the energy range 6–12 GeV. Includes in-house research effort comprising about 25 percent of the total research activity.

7. The 6-GeV proton accelerator at LBL, which is the principal source in the United States for research * using protons in the range of proton energies 1–6 GeV. Includes substantial in-house research activity.

8. A 10-GeV electron accelerator. This is a high-duty-cycle machine (in contrast to SLAC) for research * with electrons in the energy range 1–10 GeV. The in-house research activity is dominant, but there is potential for expansion to include more research by outside users.

9. A 6-GeV electron accelerator with a high duty cycle, which has recently been limited to activities associated with the development of a bypass to serve as a storage ring to study the collisions of 3.5-GeV electrons and positrons. It is the only such facility presently available in the United States and is currently under test.

10. University research groups are responsible for carrying out most of the experimental particle physics research at the major accelerator laboratories. Includes activities of professors, postdocs, graduate students, and associated technical services required to provide electronics, detection equipment, data handling and analysis systems, etc., to the extent that these aspects of the research can be mounted at the universities. Includes both experimental and theoretical physicists. User groups are partly funded under this item; they do not pay any accelerator use charges.

* Each accelerator is a source of the indicated primary particles and many beams of secondary particles (pions, K-mesons, neutrinos, muons, antiprotons, hyperons, etc.).

Nuclear Physics (Fiscal Year 1969)

Program Elements [a]	Federal Operations Support [b, c] ($ Millions)	PhD Manpower [a, d]
1. Nuclear excitation ⎫ 2. Nuclear dynamics ⎭	33.7	695
3. Heavy-ion interactions	3.1	20
4. Higher-energy nuclear physics	7.9	95
5. Neutron physics	7.7	115
6. Nuclear decay studies	3.2	65
7. Weak and electromagnetic interactions	0.7	20
8. Nuclear facilities and instrumentation	3.8	50
9. Nuclear astrophysics	0.4	20
10. Nuclear theory	5.0	260

[a] These program elements do not include all current basic research activities in nuclear physics supported by the federal government. Approximately 140 PhD's are working in the areas of data compilations, nuclear chemistry, etc.

[b] Nonfederal support estimated at 25–30 percent of federal support, on the average, of those projects supported by the federal agencies.

[c] Construction costs of major facilities not included.

[d] Another 300 PhD's are working either in applied nuclear physics or are supported entirely by nonfederal funds.

Description of Program Elements

1. The study of the nuclear degrees of freedom with a broad spectrum of nuclear probes.

2. The study of the nature of nuclear reactions.

3. The study of the now largely unknown interactions of masses of nuclear matter.

4. The study of nuclei with short-wavelength electron, proton, and mesonic probes.

5. The study of nuclear phenomena with a neutral strongly interacting probe.

6. The study of nuclear states via the decay of radioactive nuclei.

7. The study of fundamental symmetries in the nuclear domain.

8. The tools of nuclear physics.

9. Identification of which nuclear reactions are of importance and measurement of the relevant nuclear data—primarily cross sections.

10. The theoretical aspects of all the above fields and their relations to the fundamental nuclear interactions.

Atomic, Molecular, and Electron Physics (1970)

Program Elements [a]	Federal and Industrial (Operations) Support [a] ($Millions)	Estimated PhD Manpower [b]
1. Gas discharge	1.6	35
2. Electron physics	4.0	80
3. Lasers and masers	5.0	100
4. Atomic and molecular spectroscopy	4.0	80
5. Atomic, ionic, and molecular beams	6.4	130

[a] A substantial amount of activity in these program elements is supported from sources outside the AME subfield, e.g., plasma physics, space and planetary physics, electrical engineering, and chemistry. If included, this may double most of the above numbers.
[b] Based on estimated level of activity and 1970 National Register data in which a total of 1065 PhD scientists identified with AME physics.

Definition of Program Elements

1. Gas discharge including low- and medium-density plasmas.
2. Electron physics including the low-energy electron diffraction techniques, electron optics, electron–atom collisions, high-vacuum techniques, surface properties.
3. Lasers and masers including time and length standards, higher-order electromagnetic interactions, photon statistics, nonlinear spectroscopy, coherent x rays.
4. Atomic and molecular spectroscopy including positronium and muonium spectra, tests of quantum electrodynamics, optical pumping, vacuum uv, far-infrared and radio spectroscopy.
5. Atomic, ionic, and molecular beams including colliding beams, beam-foil spectroscopy, highly excited molecules and atoms.

Condensed Matter (1970)

Program Elements [a]	Estimated Federal and Industrial (Operations) Support ($Millions)	Estimated PhD Manpower [b]
1. Crystallography, etc.	9.0	150
2. Surface physics	20.5	340
3. Semiconductors	33.0	550
4. Nonelectronic aspects	18.5	310
5. Luminescence, etc.	9.5	190
6. Electronic properties of solid or molten metal	15.0	260
7. Magnetic properties	25.0	430
8. Quantum optics	9.0	150
9. High magnetic fields	3.5	60

Condensed Matter (1970)—*Continued*

Program Elements [a]	Estimated Federal and Industrial (Operations) Support ($Millions)	Estimated PhD Manpower [b]
10. Superfluidity	4.5	75
11. Slow neutron physics	8.5	75

[a] These program elements do not include all current basic research activity in condensed-matter physics.
[b] Based on estimated level of activity and 1970 National Register data in which 4160 PhD scientists identified with condensed-matter physics.

Description of Program Elements

1. Structures of crystals, including studies of atomic arrangements by neutron, electron, and x-ray diffraction techniques.
2. Includes all the properties of surfaces and thin films, crystal growth from vapor or the melt, properties of solid–solid interfaces.
3. Includes all the electronic properties of nonmetallics with small bandgaps in their pure states and having appreciable conductivity in suitably doped states.
4. Includes all the properties of defects and dislocations in crystals that are usually described without invoking the quantum-mechanical behavior of atoms. Includes plasticity, rupture, internal friction diffusion, ionic conduction, phonons, and lattice vibrations.
5. Includes band-structure calculations, optical properties, optical effects and electronic levels of impurities, and other information bearing on the electronic levels of insulating crystals.
6. Includes all the electrical and thermal conduction phenomena due to electrons, optical properties of metals, band-structure calculations, plasma oscillations, and superconductivity.
7. Includes electron paramagnetic and nuclear paramagnetic resonance work, studies of static magnetic susceptibilities, and all phenomena connected with ferromagnetism.
8. Includes lasers and masers, nonlinear optical effects, and other effects that can only be studied by laser light.
9. Experiments that are done in condensed matter with fields in excess of 120 kG.
10. Work with superfluid liquid helium.
11. Work requiring use of moderated neutrons from a pile.

Optics (1970)

Program Elements [a]	Estimated Federal and Industrial Support [b] ($Millions)	Estimated PhD Manpower [a, c] (Physicists)
1. Metrology	0.8	15
2. Optical information processing	0.9	17

Optics (1970)—*Continued*

Program Elements [a]	Estimated Federal and Industrial Support [b] ($Millions)	Estimated PhD Manpower [a, c] (Physicists)
3. Optical band communication	0.1	2
4. Optical systems, lens design, etc.	2.2	40
5. Laser-related light source	7.5	137
6. Holography and information storage	4.0	73
7. Integrated optics	0.6	11
8. Nonlinear optics	3.0	55

[a] These program elements do not include all areas of basic research in optics. It is estimated that there are another 690 PhD physicists working in areas not included in these program elements.
[b] The average annual cost per PhD does not vary widely across the program elements and is estimated to be $55,000/PhD.
[c] These estimates do not represent the magnitude of the manpower effort in the various program elements. They represent a judgment on the number of personnel from the physics section of the National Register of Scientific and Technical Personnel and do not include the large effort made by engineers, which is uniformly and properly considered optics.

Description of Program Elements

1. Metrology is the science of measurement. With lasers, very precise measurements may be made of such things as the distance to the moon, the compression of the earth in earthquake zones, and the deformation of large structures. Useful new phenomena will certainly be discovered.

2. Optical information processing is used to reduce blur in photographs, to enhance contrast, smooth out grain, sharpen edges, etc. It is also possible to use optical techniques for automatic photointerpretation and character recognition.

3. Communication at optical band frequencies permits the transmission of tremendous amounts of information wherever a beam of light can be sent. Long-distance communication through glass fibers now seems possible with modulated laser beams.

4. Modern computers and system science have made it possible to design optical systems and instruments that are optimized. Very large improvements can be made, particularly when new laser sources and solid-state receivers are included in the design.

5. Lasers can be made to have extremely high energy or power or power density. Others have very precise and steady wavelength, and still others can be tuned to different wavelengths. Each new improvement makes new techniques possible and simplifies the solution of old problems.

6. Holography is a method of storing an image or other information in a photographic film by recording the interference pattern between the signal-carrying light and a coherent reference wave. It offers potential advantages over other compact storage methods for large amounts of information.

7. A beam of light can be trapped and guided in a thin film on a solid surface, rather like electricity in a wire. It can then be manipulated by acoustical, electrical, or other optical signals for computer logic, modulation, scanning, or signaling. The combination is called integrated optics.

8. Some materials, when illuminated very intensely, give off light of doubled fre-

quency. In other cases, two beams mixed in a crystal give light of several sum and difference frequencies. Knowledge can be gained about the material, and useful devices can be built.

Acoustics (1971)

Program Elements	Estimated Federal and Industrial Support [a, b] ($Millions)	Estimated PhD Manpower [c]
1. Noise, mechanical shock, and vibration	1.8	35
2. Underwater sound	4.9	90
3. Music and architecture	0.8	15
4. Ultrasonics and infrasonics	1.1	20
5. Electroacoustics and acoustic instrumentation	1.1	20
6. Hearing, speech, and biophysical acoustics	0.8	15

[a] Based on estimated level of activity of physicists doing basic research in acoustics that leads to publishable reports. Costs/PhD across the program elements is assumed to be $55,000/year.
[b] Federal and industrial support of applied research in acoustics is estimated at $50 million.
[c] Based on estimated level of activity and 1970 National Register data in which a total of 325 PhD scientists identified with acoustics.

Description of Program Elements

1. The field of noise and noise abatement is a huge one in modern technology. It covers the sounds from jet engines, sonic boom, airflows in ducts and cooling systems, unwanted sounds of all kinds in housing and working areas. Closely related are the vibrations and shocks produced by machines. This program element has a considerable overlap with the program element of turbulence in fluid dynamics and has a strong interest in the problems of fluctation theory. In both of these areas, physics has a role to play, but the relative importance of physics research to the entire field is small, and the share of physics research will probably remain similarly small. There is still need, however, for fundamental research on the way in which particular noises arise and on their transmission through various media.

2. The study of sound propagation in water, and more specifically, seawater, has been enormously stimulated by military needs. Most of the work supported in underwater sound has been in technology rather than physics. There is strong overlap between this program element and that of oceanography. In rating both this field and that of noise, this overlap should be kept clearly in mind. Underwater sound will continue to play a significant role in the development of the field of oceanography.

3. Music includes studies of the character of musical sounds and how they are produced, both naturally and synthetically. Architectural acoustical studies are aimed at elucidating the factors that govern the acoustical character of concert halls and other structures, determining how these factors are related and how this knowledge can be translated into the design and construction of enclosures of specified acoustical characteristics.

[327]

4. The study of ultrasonic propagation in gases and liquids has long been a major component of physical acoustics. To traditional fluids, one should add the study of sound propagation in quantum liquids and in plasma. The use of Brillouin scattering to extend the frequency range of study upward, the prosecution of studies in liquid helium and plasma, and the application of our knowledge to border areas in chemistry and oceanography make this part of acoustics an especially lively one today. Of major interest has also been the contribution of this research to our understanding of relaxational phenomena and chemical kinetics. The study of sound propagation in solids is usually classified elsewhere than in acoustics. Of more purely acoustical interest in studies of physics in solids are high-accuracy velocity change measurements. Spin waves and acoustic NMR and EPR have also been studied widely. The field of nonlinear acoustics has grown out of ultrasonic propagation studies in fluids and has high promise of applications in underwater sound and biophysical acoustics. Infrasound sources include volcanoes, aerodynamic turbulence, weather frontal systems and tidal waves, and studies related to the large-scale behavior of the atmosphere, with application to clear-air turbulence detection and storm and tsunami tracking systems in this growing field of physical acoustics.

5. Represents the range of use of electrical and electronic techniques for devices that are acoustical in character and include modern stereophonic systems, acoustic pulse generation and detection, much of signal processing, and the use of computers in acoustics. The degree of involvement of physics with electroacoustics varies from time to time and depends on the particular stage of development of the devices and applications. Today, the most promising areas, from a physical viewpoint, are those of the direct production of ultrasound from electromagnetic radiation of a metal, the emission of acoustic radiation from dislocation walls in crystals, the use of heat pulses as sources on acoustic waves in the 10^{11}–10^{12}-Hz range, and acoustic thermometry. Many of these instrumentation studies are pioneering, and there is a substantial possibility of major advances in the production and use of sound.

6. While most of speech and hearing lie outside of physics, there is much that remains, such as models for speech production and analysis of the acoustic content of speech and the mechanism by which hearing takes place beyond the conversion of mechanical motions of the inner ear to nerve impulses, as well as the nonlinear behavior of the ear and its effect on hearing. Since all of hearing and speech can be classified as bioacoustics, it is convenient to particularize the rest of the field by the term *biophysical acoustics,* a field that goes beyond medical diagnosis and therapy. Biophysical acoustics overlaps with acoustic holography and includes the problem of communication of the deaf. Much of its basic thrust is in the development of ultrasonography and the use of sound waves and acoustic devices in medical treatment. As physics, the field is still small, but it is of growing significance, and the future potential is large.

Plasma and Fluids (1970)

Program Elements	Federal (Operations) Support [a] ($Millions)	Estimated PhD Manpower [b]
1. MHD power generation	1.0	20
2. Controlled fusion	30.0	410

Plasma and Fluids (1970)—*Continued*

Program Elements	Federal (Operations) Support [a] ($Millions)	Estimated PhD Manpower [b]
3. Fluid dynamics, plasmas, and lasers	24.0	475
4. Meteorology		
5. Computer modeling	8.5	210
6. Oceanography		
7. Lab and astrophysical plasma and fluids	0.5	10
8. Turbulence in fluid dynamics	3.0	65

[a] Approximately 55 percent of the scientists in plasmas and fluids are theorists. Annual support per PhD theorist is assumed to be $40,000.

[b] Based on estimated level of activity and 1970 National Register data in which 1110 PhD scientists identified with plasmas and fluids.

Description of Program Elements

1. It is possible using an intermediate state between plasmas and fluids, namely, a very-high-temperature conducting gas, to extract useful power by the flow of such a gas through a very strong magnetic field. The high-temperature gas may be the product of combustion, in which case, a very much higher temperature of combustion can and may be used as the initial starting state of a power-generating cycle. The feasibility of higher temperature in an MHD channel as compared to the limits imposed by boilers and turbine blades affords a possible significantly higher efficiency in power generation from the same fuel input, and, as a consequence, using MHD as a topping cycle affords the possibility of a significant improvement in the efficiency and simplicity of generating electrical power.

2. The goal of achieving useful power from controlled thermonuclear fusion of the heavy hydrogen isotopes requires the detailed and exhaustive understanding of the properties of high-temperature collisionless plasmas confined by various geometries of magnetic field. The thermal isolation afforded by various magnetic field configurations is limited by a complex hierarchy of instabilities whereby the high-temperature fusion plasma can escape and cool at the walls of the vessel. The understanding of these phenomena and work toward the solution of the applied goal represents the most advanced application and understanding of plasmas and of the physics of plasmas.

3. Fluid dynamics, plasmas, and lasers include the basic physical understanding of the properties of plasmas and fluids and the application of this knowledge. Because of its separate importance, turbulence has been excluded but lasers are mentioned to emphasize applications. An understanding of plasmas and fluids requires the very broadest knowledge of cooperative phenomena based on principles derived from the simplest individual particle interactions.

4. Meteorology is a specific branch of the physics of fluids because of the complexity of the water vapor, water, air, rotational centrifugal field, gravitational field of the earth–atmosphere system. Computer modeling, statistics, observation, and weather modification are the ingredients for understanding the earth's atmosphere.

5. Computer modeling of both fluids and plasmas has progressed to the state where the most complicated flow patterns, convection, partial turbulence, waves, instabilities,

[329]

and plasmas can now be modeled using finite-difference calculations on the more advanced computers. It is fair to state that the most advanced computer designs have, to a large extent, been motivated by the complexity of the modeling of fluid and plasma problems, particularly those associated with weapons design. In the future, we expect to see the problems of both controlled fusion, meteorology, and oceanography have an equal and dramatic bearing upon the evolution of computer complexity.

6. The fluid flow of the ocean is complicated by a set of constraints similar to that of the atmosphere, namely, rotation, gravitational field, and density stratification. In the case of the ocean, the thermohaline instabilities and density gradients lead to fluid-flow problems of great complexity. Oceanography in the context of fluid dynamics attempts to understand the fluid flow in the oceans due to the constraining forces as well as density gradients that lead to such exotic phenomena as the gulf stream, tides, and ocean waves. Understanding the interaction of the ocean and the atmosphere is a major objective of the physics of fluids of the earth.

7. Laboratory and astrophysical plasmas and fluids include the basic physical understanding of the properties of plasmas and fluids aside from turbulence, e.g., laminar flow, diffusion, transport coefficients, radiation properties, masers, lasers, and the application of this knowledge to the understanding of astrophysical phenomena.

8. Turbulence in fluids describes that quasi-random behavior that occurs when a highly coordinated flow breaks up into a series of partially correlated random fluctuations. In general, the fluid is characterized by the property that it may be infinitely extended quasi-statically with no restoring force. In addition, fluid turbulence exists without restoring forces; however, the one-body force that is included in fluid turbulence is gravity.

Astrophysics and Relativity

| Program Elements [a] | 1970 | | Proposed 10-Year Program | |
	Estimated Federal Support [b] ($ Millions)	Estimated PhD Manpower [c]	Annual Federal Support ($Millions)	PhD Manpower
1. Gamma-ray detectors in astronomy	0.5	5	1.5	10
2. Digitized imaging devices for optical astronomy	0.5	5	1.5	30
3. Infrared astronomy generally	1.0	—	3.0	60
4. Very large radio array	—[d]	—	10.0 [f]	75
5. Aperture synthesis for infrared astronomy	—	—	2.0	20
6. X- and γ-ray observatory	—[e]	—	40.0 [f]	75
7. Gravitational radiation	0.3	5	1.0	10
8. Neutrino astronomy	0.3	5	1.0	10

Astrophysics and Relativity—*Continued*

	1970		Proposed 10-Year Program	
Program Elements [a]	Estimated Federal Support [b] ($Millions)	Estimated PhD Manpower [c]	Annual Federal Support ($Millions)	PhD Manpower
9. Theoretical relativistic astrophysics	1.3	50	2.5	100
10. General relativity tests	0.8	10	4.0	20

[a] These program elements at present include only a small fraction of the total research activity in astrophysics. They identify areas of research that are ripe for exploration.

[b] Total annual federal support for A&R is estimated at $60 million. The cost of space-based observations amounts to about three fourths of the total federal support. University support of the field is substantial.

[c] The number of PhD's working in A&R is estimated at 300.

[d] A very large array for which design studies are complete and funding is being sought.

[e] The High Energy Astronomical Observatory in space proposed by NASA.

[f] Construction costs to be amortized over a 10-year period.

Description of Program Elements

1. Gamma-ray detectors of greatly improved sensitivity, particularly in the 0.5–30-MeV region, are essential for understanding the history of nucleosynthesis in the universe. Also needed are better means of detecting gamma rays (> 10 GeV), which may be present as a result of a variety of energetic processes in exploding objects.

2. Equipping all large telescopes with digitized imaging devices would greatly aid work in cosmology by speeding up observations by a substantial factor and by permitting electronic subtraction of atmospheric interference over a large dynamic range.

3. Infrared astronomy, still a young discipline, requires intensive development both in terms of conventional telescopes and the invention of new techniques to permit further exploration of such vast energy sources as radio galaxies and quasars.

4. There is now need for a very large radio array ($\sim$27 dishes) capable of achieving beam widths of the order of 1 sec of arc at centimeter wavelengths for studying the details of nearby bright sources with precision and for detecting faint sources out to the limits of the observable universe against the background imposed by many apparently brighter sources.

5. The technique for synthesizing a large aperture using small apertures, so successfully used in the radio range, is being tested in the infrared range using a system aimed at resolutions of 10^{-2} sec of arc or better in strong ir sources such as galactic nuclei. It is important to develop this technique to the ultimate extent possible, perhaps even to the limit imposed by the diameter of the earth (10^{-7} sec of arc).

6. Construction of a High Energy Astronomical Observatory in space for x and gamma rays would permit orders-of-magnitude improvement in sensitivity, position determination, spectral resolution, and variability measurements. Because x and gamma rays are emitted in great quantities by objects such as pulsars and quasars, it

is important to cosmology to determine whether the backgrounds of these radiations are intergalactic in origin or due to a large number of superimposed sources.

7. Recent experiments are yielding indications that gravitational radiation is emitted from astronomical sources. In view of the need to test the predictions of relativity and to identify the extreme conditions that must exist in any source capable of emitting such radiation, it is important to continue and refine such experiments.

8. The attempt to detect solar neutrinos is critically important because of its implication for the whole theory of stellar structure and evolution on which so much of astrophysics is based. It is necessary that attempts to detect solar neutrinos continue until decisive results are achieved.

9. Application of the equations of general relativity to astronomically observable objects is important to verify the correctness of the theory and clarify the basic processes that are occurring. As in all astrophysics, construction of theoretical models is the only way we have of interpreting the fragmentary information yielded by observations of relativistic objects. Therefore, in any balanced program, it is essential to increase our activity in theoretical model building in proportion to observational research.

10. Experimental tests of general relativity within the solar system have not achieved an accuracy adequate to distinguish Einstein's theory from competing theories of relativity. The advanced techniques and technology now available should enable clarification of this situation.

Physics in Chemistry (1968)

Program Elements	Estimated Federal and Nonfederal Support[a] ($ Millions)	Estimated PhD Scientists in the Physics–Chemistry Interface[b] Physicists	Chemists
1. Molecular structure and spectroscopy	60	520	650
2. Kinetics and molecular interactions	73	850	600
3. Condensed phases	64	460	825
4. Surfaces	28	290	275
5. Other	25	300	200

[a] Funding for scientists in the physics–chemistry interface area comes from sources that traditionally support physics and sources that traditionally support chemistry. The federal physics-related funds have been largely included in the funding estimates for AME and CM physics. No attempt was made to quantify the funding of chemistry in the United States. The average annual cost per PhD does not vary widely from program element to program element and is estimated to be $50,000/PhD.
[b] Based on estimated level of activity, the 1968 National Scientific Registry, and information provided by the Data Panel of the Physics Survey Committee.

Description of Program Elements

1. Includes spectroscopy of any sort (when structural information is its aim), quantum-mechanical studies of molecular structure (whether they are the phenomenological studies common to microwave and magnetic resonance studies or *a priori* studies of electronic structure), and electronic structure of solids in the context of the physics–chemistry interface.

2. The aspects of chemical kinetics in general that are considered part of the physics–chemistry interface, rather than pure chemistry, tend to involve reactions in the gas phase at all energies but concern reactions in condensed phases primarily at high energies.

3. Includes some parts of solid-state physics and chemistry and large portions of amorphous phases and polymers. Includes structure and dynamical properties of polymers; mechanical and electrical properties of liquids, glasses, and liquid crystals; luminescence and photoconductive properties of amorphous phases and molecular crystals; and some efforts toward developing devices such as liquid and plastic scintillators and amorphous switching devices.

4. Includes heterogeneous catalysis, sorption and evaporation, high-vacuum techniques, and reactions on surfaces and gas–solid interactions such as channeling.

5. Miscellaneous unclassified areas of the chemistry–physics interface.

Physics in Biology (1971)

Program Elements	Costs and Manpower
1. Molecular bases for biophysical processes 2. Neural physiology 3. Radiation phenomena 4. Thermodynamics, energy balance, and stability	The Panel found it impossible to attach manpower or funding figures to the individual elements. The number of PhD physicists doing basic research in these areas is estimated at 280

Description of Program Elements

1. A very broad category involving use of almost the entire arsenal of physics probes from x-ray crystallography through NMR and Mössbauer techniques to nanosecond fluorimetry.

2. Typical of sophisticated areas of study of macromolecular aggregates. Involves major design of new measurement techniques, computer simulation of neural behavior, study of signal-transmission characteristics of biological media.

3. Includes effects of both low- and high-level radiation on biological systems—uv to high-energy heavy particles—regeneration and repair mechanisms, and long-term effects on populations.

4. Basic questions of energy utilization and control in biological systems.

APPENDIX 4.B.   MIGRATION TRENDS WITHIN PHYSICS 1968–1970

The following table and figures display migration data for PhD physicists obtained from the 1968 and 1970 National Register of Scientific and Technical Personnel data tapes. Unless otherwise indicated, the manpower numbers shown in these figures refer to the 1970 data.

These data cover a total of 12,609 PhD respondents to the National Register questionnaire, who provided relevant input information; they represent 79 percent of all PhD respondents to the 1970 Register survey. It is estimated that approximately 85 percent of all PhD physicists in the United States are covered by the Register.

One general conclusion follows directly from Table 4.B.1. Physics PhD manpower is very much more mobile than has been commonly believed. During the period 1968–1970, about one third of all these PhD's changed their subfields of major interest and activity. This is again a strong indicator of the unity of physics; the subfield interfaces are highly permeable.

Particularly striking in the table are the 60 percent and the 33 percent increases in astrophysics and relativity and in optics PhD manpower, respectively. In the first case, this increase is a measure of the frontier challenge of the subfield and of its rapid transformation into an experimental and laboratory-based as well as a theoretical area of physics. In the second, it reflects rapid growth in activity in quantum optics. The 20 percent increase in PhD manpower shown for earth and planetary physics mirrors increasing activity in geophysics and oceanography particularly, and, more generally, increasing interest in environmental questions.

TABLE 4.B.1.   PhD Migration Data in the Physics Subfields 1968–1971 [a]

Physics Subfield	1968 PhD's	1970 PhD's	Percentage Change in Total	Number of PhD's Who Remained in 1968 Subfields	Percentage PhD's Who Remained in 1968 Subfields
Astrophysics and relativity	121	195	+60	82	42
Atomic, molecular, and electron	925	783	−15	440	56
Elementary-particle	1210	1064	−12	895	84
Nuclear	1674	1390	−17	1156	83
Fluid	367	441	+20	202	46
Plasma	458	396	−14	330	83
Condensed-matter	3759	3248	−14	2634	81
Earth and planetary	486	581	+20	323	56
Physics in biology	180	201	+12	112	56
Optics	637	848	+33	317	37
Acoustics	249	257	+3	171	67
Astronomy	658	484	−26	391	81
Astrophysics, relativity, and astronomy	779	679	−13	253	37

[a] Based on National Register responses from 12,609 U.S. PhD physicists.

Throughout the physics community, growing interest in biological problems and opportunities is indicated by the 12 percent increase in this interface.

The decreases, by some 12 percent to 17 percent, in the PhD manpower in elementary-particle physics, nuclear physics, condensed-matter physics, and atomic physics directly reflect the cutback in available support in these subfields.

Particularly noteworthy is the decrease, by 14 percent, in the plasma-physics manpower complement, which has occurred at a time when the momentum of the subfield and the potential for major scientific and technological success is high. In view of the potential societal benefits that would follow from these successes, the nation can ill afford a continuance of such trends.

This table also provides a measure of relative manpower stability in the different subfields during the 1968–1970 period—the period immediately following the sharp break in the growth of support for physics. It is somewhat surprising that the subfields show a marked grouping into

MIGRATION IN AND OUT OF PLASMA PHYSICS

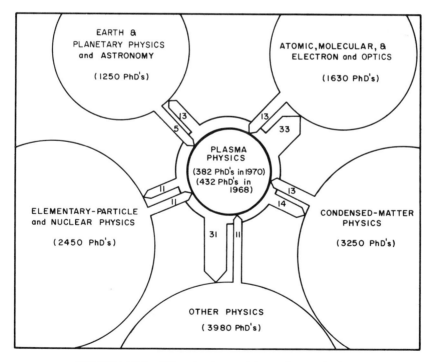

FIGURE 4.B.1   Migration into and from plasma physics.

[335]

those wherein some 80–85 percent of the PhD complement remained in its respective subfields and those in which approximately one half changed to other subfields. There is no obvious correlation in either group.

The specific interchanges between subfields are shown in Figures 4.B.1–4.B.11. It should be emphasized that the actual numbers in many cases are too small to have any statistical significance; however, the general trends indicated by these figures are of interest in establishing patterns of mobility among subfields.

MIGRATION IN AND OUT OF NUCLEAR PHYSICS

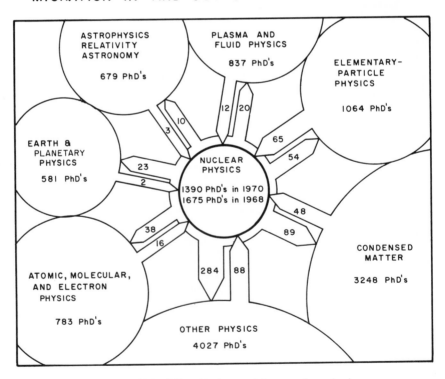

FIGURE 4.B.2   Migration into and from nuclear physics.

MIGRATION IN AND OUT OF ELEMENTARY-PARTICLE PHYSICS

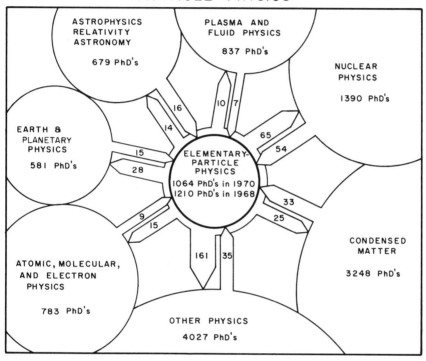

FIGURE 4.B.3   Migration into and from elementary-particle physics.

MIGRATION IN AND OUT OF ATOMIC, MOLECULAR, AND ELECTRON PHYSICS

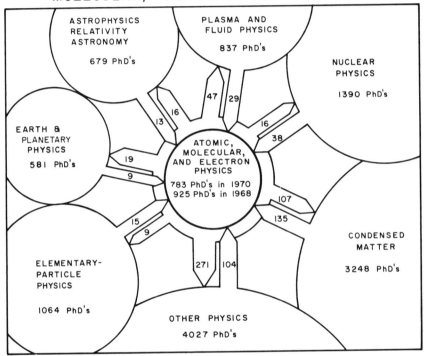

FIGURE 4.B.4   Migration into and from atomic, molecular, and electron physics.

MIGRATION IN AND OUT OF CONDENSED-MATTER PHYSICS

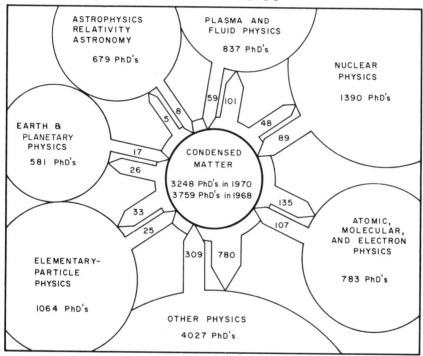

FIGURE 4.B.5 Migration into and from condensed-matter physics.

MIGRATION IN AND OUT OF ASTROPHYSICS, RELATIVITY, AND ASTRONOMY

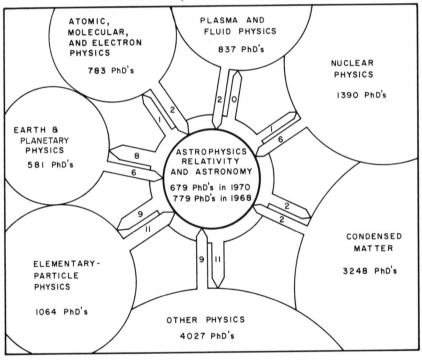

FIGURE 4.B.6 Migration into and from astrophysics and relativity and astronomy.

MIGRATION IN AND OUT OF OPTICS

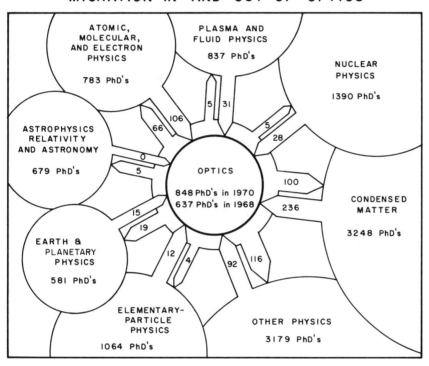

FIGURE 4.B.7   Migration into and from optics.

MIGRATION IN AND OUT OF PHYSICS
IN BIOLOGY

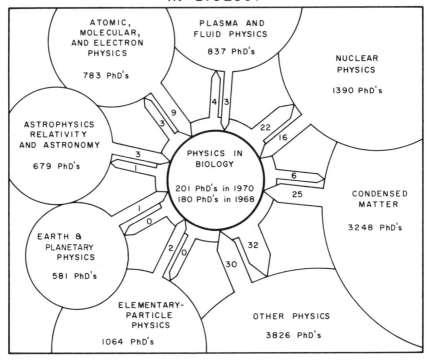

FIGURE 4.B.8   Migration into and from physics in biology.

MIGRATION IN AND OUT OF ACOUSTICS

FIGURE 4.B.9   Migration into and from acoustics.

MIGRATION IN AND OUT OF EARTH & PLANETARY PHYSICS

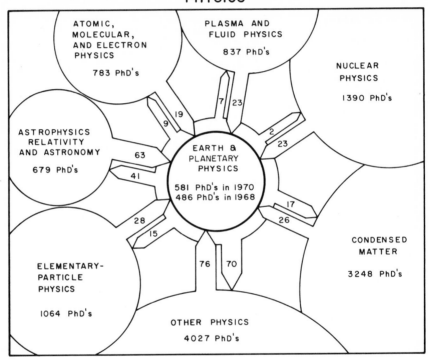

FIGURE 4.B.10   Migration into and from earth and planetary physics.

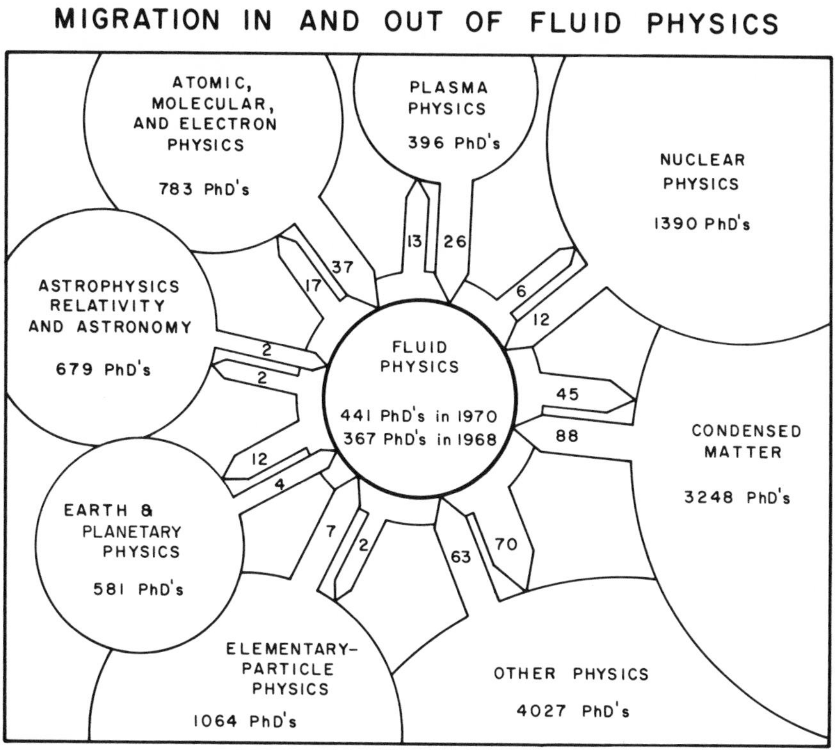

MIGRATION IN AND OUT OF FLUID PHYSICS

FIGURE 4.B.11   Migration into and from fluid physics.

APPENDIX 4.C.   SUPPLEMENTARY READING IN THE SUBFIELDS OF PHYSICS

Elementary-Particle Physics

Alvarez, L. W., "Recent Developments in Particle Physics," Science *165,* 1071 (1969).

Dyson, F. J., "Field Theory," Sci. Am. *188,* 57 (Apr. 1953).

Gell-Mann, M., and E. P. Rosenbaum, "Elementary Particles," Sci. Am. *197, 72* (July 1957).

Glaser, D. A., "The Bubble Chamber," Sci. Am. *192,* 46 (Feb. 1955).

Hearings before the Subcommittee on Research, Development, and Radiation of the Joint Committee on Atomic Energy—"High Energy Physics Research"—March 2–5, 1965.

Joint Committee on Atomic Energy. "High Energy Physics Program: Report on National Policy and Background Information." (Washington, D.C., February 1965).

Morrison, P., "The Overthrow of Parity," Sci. Am. *196,* 45 (Apr. 1957).

Sachs, R. G., "Can the Direction of Flow of Time Be Determined?" Science *140,* 1284 (1963).

[345]

Segrè, E., and C. E. Wiegand, "The Antiproton," Sci. Am. *194,* 37 (June 1956).

Treiman, S. B., "The Weak Interactions," Sci. Am. *200,* 72 (Mar. 1959).

Weisskopf, V. F., "Physics in the Twentieth Century," Science *168,* 923 (1970).

Wigner, E. P., "Violations of Symmetry in Physics," Sci. Am. *213,* 28 (Dec. 1965).

Yuan, L. C. L., ed., *Elementary Particles and Society* (Academic Press, New York, 1971).

Yuan, L. C. L., ed., *Nature of Matter, Purposes of High Energy Physics,* Brookhaven National Laboratory (1964). Available from National Technical Information Service, U. S. Department of Commerce, Springfield, Virginia 22151.

Nuclear Physics

Baranger, M., and E. W. Vogt, eds., *Advances in Nuclear Physics* (Plenum Press, New York, 1968–). An annual series of review volumes providing authoritative treatments of contemporary nuclear research.

Bromley, D. A., "The Nucleus," Phys. Today *21*(5), 29–36 (1968). A 20-year survey review of nuclear physics.

Crewe, A. V., and J. J. Katz, *Nuclear Research U.S.A.—Knowledge for the Future* (Dover, New York, 1969). An elementary survey of U.S. nuclear research activity, profusely illustrated.

Cohen, B. L., *Concepts of Nuclear Physics* (McGraw-Hill, New York, 1971). An intermediate survey of nuclear structure and reaction research.

Cohen, B. L., *The Heart of the Atom* (Doubleday, New York, 1967). An elementary survey of nuclear-structure studies.

Mottelson, B. R., and A. Bohr, *Nuclear Structure* (Benjamin, New York, 1969). A three-volume advanced treatise on nuclear physics.

Seaborg, G. T., and W. R. Corliss, *Man and Atom: Building a New World Through Nuclear Technology* (Dutton, New York, 1971).

Atomic, Molecular, and Electron Physics

Bates, D. R., and I. Estermann, eds., *Advances in Atomic and Molecular Physics* (Academic Press, New York, 1965 to present), Vols. 1–5.

Levine, A. K., ed., *Advances in Lasers* (Academic Press, New York, 1966 to present), Vols. 1–4.

Marton, L., ed., *Advances in Electronics and Electron Physics* (Academic Press, New York, 1948 to present), Vols. 1–27 and Supplements 1–7.

* Ramsey, N. F., *Molecular Beams* (Oxford U.P., New York, 1956).

Schawlow, A. L., ed., *Lasers and Light.* Readings from the *Scientific American* (Freeman, San Francisco, 1969).

* Schawlow, A. L., and C. H. Townes, *Microwave Spectroscopy* (McGraw-Hill, New York, 1955).

Yariv, A., *Quantum Electronics* (Wiley, New York, 1967). An intermediate textbook with an account of masers, lasers, and nonlinear optics.

---

* These references are somewhat older, standard classics that provide a good introduction to modern atomic and molecular spectroscopy. Basic textbooks in quantum theory, statistical mechanics, optics, and electromagnetic radiation are both numerous and essential for a good grasp on this subfield.

Condensed Matter

Alexander, W. O., "The Competition of Materials," Sci. Am. *217*, 254 (Sept. 1967).

Charles, R. J., "The Nature of Glasses," Sci. Am. *217*, 126 (Sept. 1967).

Committee of the Solid State Sciences Panel, *Research in Solid-State Sciences: Opportunities and Relevance to National Needs,* Publ. 1600 (National Academy of Sciences, Washington, D.C., 1968).

Cottrel, A. H., "The Nature of Metals," Sci. Am. *217*, 90 (Sept. 1967).

Gilman, J. J., "The Nature of Ceramics," Sci. Am. *217*, 112 (Sept. 1967).

Ehrenreich, H., "The Electrical Properties of Materials," Sci. Am. *217*, 194 (Sept. 1967).

Ehrenreich, H., F. Seitz, and D. Turnbull, eds., *Solid State Physics* (Academic Press, New York), Vol. I, 1955–Vol. XXVI, 1971.

Huggins, R. A., R. H. Bube, and R. W. Roberts, eds., *Annual Review of Materials Science* (Annual Reviews, Palo Alto, Calif., 1971), Vol. 1.

Javan, A., "The Optical Properties of Materials," Sci. Am. *217*, 238 (Sept. 1967).

Keffer, R., "The Magnetic Properties of Materials," Sci. Am. *217*, 222 (Sept. 1967).

Kelly, A., "The Nature of Composite Materials," Sci. Am. *217*, 160 (Sept. 1967).

Kittel, C. *Introduction to Solid State Physics* (Wiley-Interscience, New York, 1971), 4th ed.

Mark, H. F., "The Nature of Polymeric Materials," Sci. Am. *217*, 148 (Sept. 1967).

Mott, N., "The Solid State," Sci. Am. *217*, 80 (Sept. 1967).

Reiss, H., "The Chemical Properties of Materials," Sci. Am. *217*, 210 (Sept. 1967).

Smith, C. S., "Materials," Sci. Am. *217*, 68 (Sept. 1967).

Ziman, J., "The Thermal Properties of Materials," Sci. Am. *217*, 180 (Sept. 1967).

Optics

Born, M., and E. Wolf., *Principles of Optics* (Macmillan, New York, 1964), 2nd revised ed. An exhaustive and highly mathematical treatment of all optical phenomena that may be treated by classical electromagnetic theory. It is a definitive reference work in optics.

Collier, R. J., C. R. Burckhardt, and L. L. Lin, *Optical Holography* (Academic Press, New York, 1971). This text begins with basic mathematics and optical concepts and then proceeds through nearly all aspects of the theory and art of holographic image formation with visible light.

DeVelis, J. B., and G. O. Reynolds, *Theory and Applications of Holography* (Addison-Wesley, Reading, Mass., 1967). Authoritative and well illustrated. Contains much advanced mathematics.

Goodman, J. W., *Introduction to Fourier Optics* (McGraw-Hill, New York, 1968). The work is directed toward an understanding of the applications of Fourier analysis and linear systems concepts to optics. It is particularly directed toward those familiar with network analysis but relatively weak in the principles of classical optics.

Jenkins, F. A., and H. E. White, *Fundamentals of Optics* (McGraw-Hill, New York, 1957), 3rd ed. A renowned college text by two distinguished teachers. It is especially recommended for its treatment of classical physical optics.

O'Neill, E. L., *Introduction to Statistical Optics* (Addison-Wesley, Reading, Mass., 1963). An introduction to classical statistical optics. It is suitable source material for seniors and first-year graduate students in physics or electrical engineering.

[347]

Schawlow, A. L., ed., *Lasers and Light*. Readings from the Scientific American (Freeman, San Francisco, 1969). This book contains 11 offprints of articles from the *Scientific American,* September 1968, and 21 offprints from earlier editions, making a splendid collection on light for college and high-school students and for the general reader.

Smith, W. J., *Modern Optical Engineering* (McGraw-Hill, New York, 1966). This book is intended for the individual whose background is in engineering or physics. It is designed to provide the information necessary for such an individual to pursue a career in optical systems design.

Wolf, E., ed., *Progress in Optics* (North-Holland, Amsterdam), Vol. I, 1961–Vol. IX, 1971. This series, which publishes an annual volume, contains review articles about current research in optics and related fields. Its purpose is to help optical scientists and engineers be well informed about advances in the field.

Wood, R. W., *Physical Optics* (Macmillan, New York, 1934), 3rd ed. Still unique is this entertainingly written and emphatically experimental description of the phenomena of physical optics, from which the mathematics of the electromagnetic theory was progressively removed as successive editions appeared.

Acoustics

Backus, J. M., *The Acoustic Foundations of Music* (Norton, New York, 1969). A look at the elements of music from the viewpoint of a physicist.

Békésy, G. V., *Experiments in Hearing* (McGraw-Hill, New York, 1960). An edited compilation of Békésy's published articles. Deals with physiological as well as with psychological aspects of hearing. Contains work that earned Békésy the 1961 Nobel Prize in Physiology and Medicine.

Beyer, R. T., and S. V. Letcher, *Physical Ultrasonics* (Academic Press, New York, 1969). A detailed discussion of prominent areas of ultrasonic research.

Flanagan, J. L., *Speech Analysis, Synthesis and Perception* (Academic Press, New York, 1965). A study of the acoustical properties of the voice and ear, techniques of speech analysis and synthesis, with mathematical accounts of models and systems.

Green, D. M., and J. A. Swets, *Signal Detection Theory and Psychophysics* (Wiley, New York, 1966). Fundamental text for much of modern psychophysics, especially psychoacoustics. The authors belong to a small group of theoreticians who adapted the general theory of signal detectability to psychophysics.

Harris, C. M., ed., *Handbook of Noise Control* (McGraw-Hill, New York, 1957). Articles on numerous aspects of noise, its effect on man and its control, written by more than a score of the leading workers in the field.

Harris, C. M., and C. E. Crede, *Shock and Vibration Handbook* (McGraw-Hill, New York, 1961). A three-volume set covering practical shock and vibration problems.

Kinsler, L. E., and A. R. Frey, *Fundamentals of Acoustics* (Wiley, New York, 1962), 2nd ed. The leading general text in acoustics.

Knudsen, V. O., and C. M. Harris, *Acoustical Designing in Architecture* (Wiley, New York, 1950). A classic in its field by two of the outstanding practitioners of auditorium design.

Kryter, K. D., *The Effects of Noise on Man* (Academic Press, New York, 1970). An extensive compilation of current knowledge concerning psychological and physiological effects of noise on people. Deals with fundamental and applied aspects of the problem.

Mason, W. P., ed., *Physical Acoustics* (Academic Press, New York, 1964–1970), Vols. 1–7. A comprehensive, continuing series on the physical side of acoustics. Now in seven volumes, many of which consist of two books.

Morse, P. M., and K. U. Ingard, *Theoretical Acoustics* (McGraw-Hill, New York, 1968). An exhaustive mathematical treatise, covering primarily vibrations, sound radiation, and scattering.

Truell, R., C. Elbaum, and B. B. Chick, *Ultrasonic Methods in Solid State Physics* (Academic Press, New York, 1969). A detailed presentation of the basic methods used in the investigation of the solid state by acoustical means.

Urick, R. J., *Principles of Underwater Sound for Engineers* (McGraw-Hill, New York, 1967).

Plasma and Fluid Physics

Batchelor, G. K., *The Theory of Homogeneous Turbulence* (The University Press, Cambridge, England, 1960).

Bishop, A. S., *Project Sherwood* (Addison-Wesley, Reading, Mass., 1958).

Courant, R., and K. O. Friederichs, *Supersonic Flow and Shock Waves* (Interscience, New York, 1948).

Glasstone, S., and R. H. Lovberg, *Controlled Thermonuclear Reactions* (Van Nostrand, Princeton, N. J., 1960).

Grad, H., "Frontiers of Physics Today: Plasmas," Physics Today 22, 34 (Dec. 1969).

Lamb, H., *Hydrodynamics* (Dover, New York, 1945).

Leslie, D. C., *Developments in the Theory of Turbulence* (Oxford U. P., Oxford, England, in press).

Lessing, L., "New Ways to More Power With Less Pollution," Fortune 82, 78–81 (Nov. 1970).

Oswatitsch, K., *Gas Dynamics* (Academic Press, New York, 1956).

Rosa, R. J., "Physical Principles of Magnetohydrodynamic Power Generation," Phys. Fluids 4, 182–194 (1961).

Spitzer, L., *Physics of Fully Ionized Gases* (Interscience, New York, 1962), 2nd ed.

Tennekes, H., and J. L. Lumley, *A First Course in Turbulence* (MIT Press, Cambridge, Mass., 1972).

Astrophysics and Relativity

Calder, N., *Violent Universe* (Viking, New York, 1969).

Gamow, G., *Creation of the Universe* (Viking, New York, 1952).

Frontiers in Astronomy, a collection of *Scientific American* articles (Freeman, San Francisco, 1970).

Menzel, D. H., F. L. Whipple, G. de Vaucouleurs, *Survey of the Universe* (Prentice-Hall, Englewood Cliffs, N. J., 1970).

Schatzman, E. L., *The Structure of the Universe* (World University Library, New York, 1968).

Earth and Planetary Physics

Commission on Marine Science, Engineering, and Resources, *Our Nation and the Sea: A Plan for National Action* (U.S. Govt. Printing Office, Washington, D.C., 1969).

Committee on Atmospheric Sciences, *The Atmospheric Sciences and Man's Needs: Priorities for the Future* (National Academy of Sciences, Washington, D.C., 1970).

Committee on Atmospheric Sciences, *Weather and Climate Modification: Problems and Prospects,* Publ. 1350 (National Academy of Sciences–National Research Council, Washington, D.C., 1966).

Committee on Earthquake Engineering Research, *Earthquake Engineering Research* (National Academy of Sciences, Washington, D.C., 1969).

Committee on Oceanography and Committee on Ocean Engineering, *An Oceanic Quest: The International Decade of Ocean Exploration* (National Academy of Sciences, Washington, D.C., 1969).

Committee on Seismology, *Seismology: Responsibilities and Requirements of a Growing Science* (National Academy of Sciences, Washington, D.C., 1969).

Committee on Solar-Terrestrial Research of the Geophysics Research Board, *Physics of the Earth in Space: The Role of Ground-Based Research* (National Academy of Sciences, Washington, D.C., 1969).

Cressman, G. P., "Public Forecasting—Present and Future," in *A Century of Weather Progress* (Am. Meteorol. Soc., Boston, Mass., 1970), pp. 71–77.

Division of Engineering, *Useful Applications of Earth-Oriented Satellites* (National Academy of Sciences-National Research Council, Washington, D.C., 1969).

National Aeronautics and Space Administration, *The Terrestrial Environment: Solid-Earth and Ocean Physics,* NASA CR-1579 (NASA, Washington, D.C., 1970).

Space Science Board, *Priorities for Space Research 1971–1980* (National Academy of Sciences, Washington, D.C., 1971).

Physics in Chemistry

Cooper, L. N., *An Introduction to the Meaning and Structure of Physics* (Harper and Row, New York, 1968).

Eyring, H., ed., *Annual Reviews of Physical Chemistry* (Annual Review, Inc., Palo Alto, California, 1950–).

Hinshelwood, C. N., *The Structure of Physical Chemistry* (Cambridge U. P., New York, 1961).

Prigogine, I., ed., *Advances in Chemical Physics* (Wiley, New York, 1958–).

Young, L. B., ed., *The Mystery of Matter.* The American Foundation for Continuing Education (Oxford U. P., New York, 1965).

Physics in Biology

Benzer, S., "Adventures in the rII Region," in *Phage and the Origins of Molecular Biology,* J. Cairns, G. S. Stent, and J. D. Watson, eds. (Cold Spring Harbor Laboratory of Molecular Biology, 1966).

Dyson, F. J., "The Future of Physics," Phys. Today *23,* 23 (Sept. 1970).

Goldman, L., and R. J. Rockwell, *Lasers in Medicine* (Gordon & Breach, New York, 1970).

Olby, R., "Francis Crick, DNA, and the Central Dogma," Daedalus *99,* 938–987 (1970).

Schrodinger, E., *What is Life?* (Cambridge U. P., Cambridge, Mass., 1944), reprinted 1967.

Wilkins, M. H. F., "Molecular Configuration of Nucleic Acids," Science *140,* 941 (1963).

APPENDIX A
PHYSICS
SURVEY
COMMITTEE

D. ALLAN BROMLEY, Yale University, *Chairman*
*DANIEL ALPERT, University of Illinois
RAYMOND BOWERS, Cornell University
JOSEPH W. CHAMBERLAIN, The Lunar Science Institute
HERMAN FESHᴿACH, Massachusetts Institute of Technology
GEORGE B. FIELD, University of California, Berkeley
RONALD GEBALLE, University of Washington
CONYERS HERRING, Bell Telephone Laboratories, Inc.
ARTHUR R. KANTROWITZ, AVCO-Everett Research Laboratory
THOMAS LAURITSEN, California Institute of Technology
FRANKLIN A. LONG, Cornell University
E. R. PIORE, International Business Machines Corporation
E. M. PURCELL, Harvard University
ROBERT G. SACHS, University of Chicago
*CHARLES H. TOWNES, University of California, Berkeley
ALVIN M. WEINBERG, Oak Ridge National Laboratory
VICTOR F. WEISSKOPF, Massachusetts Institute of Technology

GEORGE W. WOOD, *Executive Secretary*

* Members of original Survey Committee. Other responsibilities precluded these
physicists from participating in other than the early deliberations.

[351]

APPENDIX B
CONTENTS
Physics in Perspective
Volume I

[353]